天津·国家海洋博物馆

Tianjin · National Maritime Museum of China

刘景樑　主编

By Liu Jingliang

天津大学出版社

TIANJIN UNIVERSITY PRESS

谨以此书献给
为国家海洋博物馆项目作出贡献的人们

Dedicate this book to
people who have contributed to the National Maritime Museum of China

众出口
AUDIENCE EXIT
参观请直行
龙的时代
远古海洋
今日海洋

潮汐预报
潮流预报
国家海洋博物
National Maritime Museum of C

序一 海的殿堂

Preface I
A Hallowed Hall of Sea

单霁翔（中国文物学会会长）

Shan Jixiang (President of the Chinese Society of Cultural Relics)

社会文明进步的标志之一是人类对赖以生存之自然与文化的敬畏。国家海洋博物馆创新的建筑造型、丰富的自然与人文藏品，使它不仅是一处保护之所，更是一座有非凡的遗产教育功能的知识殿堂。

在我记忆中，该馆的兴建源自 2007 年几十位院士的联名提议。2013 年，天津市建筑设计院（TADI）刘景樑大师领衔的团队与澳大利亚 COX 建筑师事务所的合作方案在终审中一举夺魁，而后我被推选为国家海洋博物馆专家委员会主任，一晃六七年过去了。2019 年 5 月，该馆面向公众试运行，并获广泛好评。今天，由刘景樑大师主编的图文并茂的《天津 · 国家海洋博物馆》样书在《中国建筑文化遗产》编辑部的鼎力支持下呈现在我面前。翻阅着彰显刘景樑大师团队设计与营造技艺的学术成果，我很钦佩，并由衷地感到国家海洋博物馆是天津乃至中国充满创新意义的标志性文化建筑。从博物馆建筑的多元价值上看，其可贵之处在于它具备文化、艺术、生态和人文的多重属性，当地政府可以通过博物馆提升城市形象与竞争力，吸引更多公众造访。生机盎然的展馆造型与展览空间，在彰显建筑个性的同时，更让观众在知识的海洋里流连忘返，留下更加美好的记忆。

国家海洋博物馆专家委员会先后组织了一系列研讨与论证活动。2014 年 9 月 14 日，第二次专家委员会会议在故宫博物院召开，专家们先后听取了建筑与策展设计思路介绍，提出了进一步合理规划参观流线、充分考

One mark of the progress in civilization is the awe of nature and culture on which human beings depend for survival. With its innovative architectural style and rich natural and cultural collections, the National Maritime Museum of China has become not only a place of conservation, but also a hallowed hall of knowledge with the extraordinary function of heritage education.

I remember that the idea of building the museum was jointly proposed by dozens of academicians in 2007. In 2013, the team spearheaded by Master Liu Jingliang of Tianjin Architectural Design Institute and COX Architecture of Australia won the final award, and then I was elected the Director of the Expert Committee of the National Maritime Museum of China. Six or seven years have passed by swiftly. Since it was put into trial operation and open to the public in May 2019, the museum has been widely acclaimed. Today, completed with the full support of the editorial department of *China Architectural Heritage*, the sample of *Tianjin·National Maritime Museum of China*, an illustrated book edited by Master Liu Jingliang is presented to me. When reading the academic achievements that show the design and construction skills of Master Liu Jingliang's team, I marvel at the National Maritime Museum of China which is an innovative cultural landmark in Tianjin and even China. The multiple values of the museum architecture are reflected in its diverse attributes of culture, art, ecology and humanity. The local government can enhance the image and competitiveness of the city through the museum, attracting more public visitors. The vibrant forms and space of the exhibition halls not only present the individuality of the architecture, but also draw visitors' attention, arousing their interest in the ocean of knowledge and leaving a wonderful memory in their mind.

The Expert Committee of the National Maritime Museum of China has held a series of discussions and demonstration activities. On September 14, 2014, the second meeting of the Expert Committee was held in the Palace Museum, in which experts listened to the introduction of ideas for architectural and

虑节能环保因素、在展陈等的策划中增加中华海洋文明自信和海洋文明复兴等方面的建议。据我所知，现在建成的国家海洋博物馆不仅对此加以优化与完善，还特别注重参观动线和展陈效果的设计，以激发参观者的兴趣，更依靠生动的展陈吸引公众的关注。我记得 2017 年 7 月 21 日，第四次专家委员会在故宫博物院召开，重点研讨了国家海洋博物馆一期开放的中华海洋文明、海上丝绸之路、远古海洋、今日海洋展区及其对实施海洋启蒙有直接作用的儿童教育中心和全景沉浸式影院等设施的深化设计成果。通过建筑与文博专家们的通力合作，国家海洋博物馆的建设仿佛翻开文化建筑新篇章，展现出山海形胜、文化之邦的气概。恰恰由于各方设计团队从一开始就遵循了复合化的文化建筑策略，从而在理性与感性交织上、在艺术与科技结合上探索了新路。我认为，一个好的博物馆建筑需要贯穿始终的优秀设计，既要有超越的文化意象，还要有与城市设计共通的科学统筹，更离不开“形式追随功能”的设计内核。国家海博馆之所以让观者印象深刻，既在于它的展陈新颖丰富，也在于它的建筑塑形充满场所想象与情怀。设立在海边的博物馆入口空间通过海洋的烘托，既有稳重静谧的气质，也轻盈灵动地似艺术拱桥一般欢迎着四海之宾的到来，让人仿佛步入了海上丝绸之路的“乐园”。

我认为，博物馆成为“精神的家园”，首先在于博物馆要体现社会责任与城市文化。作为城市文化的重要

curatorial design, and put forward their suggestions on further rational planning of visiting streamline, full consideration of energy conservation and environmental protection, boosting confidence in Chinese maritime civilization in exhibition, rejuvenation of maritime civilization and so on. As far as I know, the museum just built has not only optimized these ideas, but also paid special attention to the design of visitors' movement patterns and exhibition effects to stimulate their interest, and attract the public attention by presenting vivid exhibitions.

I remember that the fourth meeting of the Expert Committee was held on July 21, 2017 in the Palace Museum, focusing on the design achievements of the areas exhibiting Chinese maritime civilization, the Maritime Silk Road, ancient oceans, today's oceans, children's education center with a direct effect on the maritime enlightenment, and a panoramic immersive cinema, which were all launched at Phase I of the National Maritime Museum of China. With the cooperation of architectural and cultural experts, the construction of the museum has opened a new chapter in cultural architecture, showing the wonders of mountains and seas and the prosperity of culture in our country. As the design teams of all parties have followed the strategy of compound cultural architecture from the very beginning, they explored new ways in the interweaving of sense and sensibility and the combination of art and technology. In my opinion, a good museum building calls for excellent design in every aspect, including transcendent cultural images, scientific coordination with urban design, and an inalienable design core of "forms following functions". The National Maritime Museum of China leaves a deep impression on its visitors not only by its novel and rich exhibitions but also its architectural forms full of spatial imaginations and noble sentiments. The entrance space of the museum at seaside stands solemnly and tranquilly against the backdrop of a sea, and expands lightly and vibrantly like an arch bridge of art, welcoming the arrival of guests from all over the world. Visitors may feel they are stepping into a

载体，博物馆见证并凝聚着城市发展的历程，在城市记忆的保持、特色形象的展示、乡土情结的维系、文化身份的认同等方面体现综合价值。恰恰为此，一个好的博物馆建筑就必须在反映城市文脉、表现国家文化上体现新创意和新举措，并使馆藏和展陈的文化与价值深植于建筑形态之中。2012 年，我参加了中国建筑学会主办的“建筑方针 60 年的当代意义”研讨会，我从创造精品博物馆建筑入手，归纳了博物馆建筑设计应关注的八个“回归”，即博物馆建筑应回归城市理想、回归历史责任、回归永恒价值、回归文化特征、回归科学精神、回归社会期待、回归生态环境、回归服务职能。今天，国家又确认了“适用、经济、绿色、美观”的建筑新“八字方针”。对比之下，我确信国家海洋博物馆的建筑体现了建筑师在大海洋观理念下独有的设想、计划与构思，它使设计思维创意变成现实，是建筑师思想升华的结果。国家海洋博物馆是具有恒久生命力的建筑，文化理想赋予它更多的审美内涵，历史责任也让它拥有胸怀与灵魂，这让它的功能超越一般建筑的使用功能，必然会拥有经久不衰的生命力。此外，国家海洋博物馆是学术性与公众性兼具的国家专业大馆，其具有国际视野、反映大国海洋宝藏的展陈，全方位、立体化服务中外参观者的形式，都使它获得越来越高的赞誉，影响力不断提升，使它不愧为一座具有世界领先水准的国家级海洋文化博览建筑。

"wonderland" of the Maritime Silk Road.

In my opinion, to become a "spiritual home", a museum should first embody social responsibility and urban culture. As an important carrier of urban culture, it witnesses the course of urban development and presents that in a miniature, reflecting comprehensive values in terms of maintaining urban memory, displaying featured images, retaining local complex, and showcasing cultural identity. For this reason, a good museum building must embody new ideas and measures in reflecting the culture of a city and expressing national culture, and have the culture and value of collections and exhibitions deeply rooted in the architectural form. In 2012, I attended the "Seminar on the Contemporary Significance of Architectural Guidelines of Six Decades" sponsored by the Architectural Society of China. Starting with creating excellent museum buildings, I summarized eight "returns" that should be noted in museum architectural design, namely, museum buildings should return to urban ideals, historical responsibilities, eternal values, cultural characteristics, scientific spirits, social expectations, ecological environment and service functions. Today, our state has confirmed the new architectural guidelines of "applicability, economy, green and beauty". I am sure that the building of the National Maritime Museum of China embodies the architects' unique vision, plan and conception guided by the macro maritime outlook, materializes their creativity and sublimates their thoughts. The museum is a building with permanent vitality because its cultural ideals render it more aesthetic connotation, and its historical responsibilities endow it with a mind and soul, making it surpass the functions of ordinary buildings. It is bound to endure. In addition, as a national specialized museum that combines academic research and service for the general public, the National Maritime Museum of China hosts exhibitions that, from an international perspective, reflectting maritime treasures of a major power, and serving Chinese and foreign visitors in an all-round way. All of the above have made it gain higher acclaim

刘景樑大师及其团队以倾心近十载的设计成果为基础写作的《天津·国家海洋博物馆》一书即将面世，我认为该项目的设计文化表达无论从“创造性”，还是从“力量性”都为天津乃至中国海洋文化建设填补了“空白”，并为启迪国民走向蓝色国土、建设可持续海洋文明，提供了建筑师贡献的完满“样本”。在此我感谢并祝贺刘景樑大师与《中国建筑文化遗产》编辑部等共同完成的《天津·国家海洋博物馆》一书的出版，特以此文为序。

2020 年 8 月

as well as increasing influence, and have become a world-class museum for maritime culture exhibitions.

The book *Tianjin·National Maritime Museum of China* will be published soon. It is written by Master Liu Jingliang and his team based on the design achievements of nearly one decade. I think that the book as an expression of the design culture of this project has filled a gap for the development of maritime culture in Tianjin and even China from the perspective of both "creativity" and "power". It can be deemed a perfect architectural work that kindles people's interest in blue oceans and helps build a sustainable maritime civilization. Here, I would like to thank and congratulate Master Liu Jingliang and the editorial department of *China Architectural Heritage* for completing the publication of *Tianjin·National Maritime Museum of China*, and write this article as a preface to the book.

August 2020

序二　向海而生

Preface II Born Facing the Sea

路　红（天津市规划和自然资源局 一级巡视员）

Lu Hong (Class I Inspector of the Tianjin Municipal Bureau of Planning and Natural Resources)

2018 年 4 月，习近平总书记在海南考察时指出“我国是一个海洋大国，海域面积十分辽阔。一定要向海洋进军，加快建设海洋强国”。建设海洋强国的一个重要的基础工作就是要增强中华民族的海洋意识，让人们认识海洋，重塑海洋价值观。坐落在天津滨海新区中新天津生态城的国家海洋博物馆就承担了这个重任，她是首座国家级的以海洋文化为主题的综合性、公益性博物馆，丰富的展陈内容将使其成为人类与海洋互动的新场所，更是中国海洋文化事业的里程碑。

国家海洋博物馆经过长期周密的筹划和历时 6 年的建设，于 2019 年落成。国家海洋博物馆的建设从筹备之初，就得到了自然资源部（含原国家海洋局）和天津市委、市政府的关怀指导，得到了天津市滨海新区政府的支持，也凝聚了在筹划、设计、施工、管理各阶段各级领导、专家和建设者付出的心血。天津市规划和自然资源局直属的天津市建筑设计院、国家海洋博物馆筹建办作为设计和管理团队，从建筑设计到展览布置、开馆筹备等，均付出了极大的努力，也取得了丰硕的成果。国家海洋博物馆自 2019 年 5 月开放至 2019 年底，接待了来自国内外的 130 万人次以上的参观者，其丰富多彩的展陈内容、独具特色的建筑形象和内外交融的海陆环境，得到了国内外观众的交口称赞。如今，她已成为海洋强国的宣教科普阵地

In April 2018, General Secretary Xi Jinping pointed out that China has a major maritime power with a vast sea area. We must march into the ocean and speed up the construction of our country as a maritime power. during his visit in Hainan. An important basic mission for building a maritime power is to enhance the Chinese nation's maritime awareness, let people know about the ocean, and reshape their maritime values. The National Maritime Museum of China standing in the China-Singapore Tianjin Eco-city of Tianjin Binhai New Area undertakes this important mission. As the first state-level comprehensive and public museum on the theme of maritime, this museum with diversified collections, will become a new place for human interaction with the ocean, and mark a milestone in China's maritime culture.

The National Maritime Museum of China was completed in 2019 after a long period of careful planning and six years of construction. From the beginning of preparation, it has received the care and guidance from the Ministry of Natural Resources (State Oceanic Administration), the Tianjin Municipal Party Committee and Municipal Government, and the support of Binhai New Area Government. The project also embodies the painstaking efforts of officials, experts and builders at all stages and levels of planning, design, construction and management. As the design and management teams, Tianjin Architectural Design Institute directly under Tianjin Municipal Bureau of Planning and Natural Resources and the National Maritime Museum Preparatory Office have made great efforts, achieving fruitful results from architectural design to exhibition layout and opening preparation. From its opening in May 2019 to the end of 2019, the museum received over 1.3 million visitors from home and abroad. Its diversified exhibition contents, unique architectural image and the sea (outdoors) and land (indoors) environment have been unanimously praised by both domestic and foreign visitors. Now, the beautiful museum has become a center for publicity and education on China's maritime power and a new

和天津靓丽的新地标、新名片。

当代建筑学观念认为"建筑学是为人类建立生活环境的综合艺术和科学。建筑师的责任是把已有的和新建的、自然的建筑与人为的因素结合起来，并通过'用设计满足人类需求的建筑'改变城市面貌。建筑师应继承和发展社会遗产，为社会创造新的建筑形式，并保持文化发展的连续性"（《建筑师的华沙宣言》，1981 年，国际建筑师协会第 14 次世界建筑师大会发布）。据此观点，我谨从一名建筑师的角度，从三个方面来评价国家海洋博物馆的建筑设计。一是为人类服务的功能突出合理。国家海洋博物馆是集收藏保护、展示教育、科学研究、文化传播、休闲旅游于一体的综合性博物馆建筑，建筑师将当代科技和博物馆的功能结合设计，为观众创造了丰富的海洋类展品的展览空间、教育空间、研究空间和休憩空间，创造了宽敞连续的空间和流畅的参观路线：从主入口步入宽敞的中央大厅，以此为交通主枢纽，在两侧为观众提供报告厅、电影厅、餐厅、商店等服务设施，中央大厅北端向东、西延伸，与一至四号展厅有机串联，形成了有序的展览空间。其他辅助空间、观众的休憩空间、研究和工作空间各得其所。二是建筑与环境交融。国家海洋博物馆作为地标性建筑，在结合环境地形、创造临海景观的可视性方面做了很多巧妙的设计。在室外，环绕配合五指形的建筑主体，形成了各具功能

landmark and calling card of Tianjin.

According to the contemporary architectural concept, "architecture is a comprehensive art and science for building a living environment for human beings. The architects' responsibility is to combine the existing and newly built, natural and artificial factors, and to change the city's appearance by 'the building that meets human needs with design'. Architects should inherit and develop social heritage to create new architectural forms for society and maintain the continuity of cultural development" (*Warsaw Declaration of Architects*, 1981, issued by the 14th World Architects Congress of the International Union of Architects). From this point of view, I would like to evaluate, as an architect, the architectural design of the National Maritime Museum of China from three aspects. First of all, the function of serving mankind is outstanding and reasonable. As the museum is a comprehensive building integrating collection, conservation, exhibition, education, scientific research, cultural communication and leisure tourism, the architects have combined contemporary science and technology with the functions of the museum to create diversified exhibition spaces, educational spaces, research spaces and rest spaces for visitors, and create spacious and continuous spaces and smooth visiting routes. Visitors can walk from the main entrance into the spacious central hall, which serves as the main transportation hub, with service facilities on both sides for visitors, such as a lecture hall, a movie hall, restaurants and shops. The northern end of the central hall extends eastward and westward to organically connect with exhibition halls 1–4, forming orderly exhibition spaces. Other auxiliary spaces, visitors' rest spaces, research and work spaces are all properly arranged.

Second, the architecture and the environment are integrated. Many ingenious designs in the museum as a landmark building have been done by combining the environment and terrain and creating the visibility of coastal view. Outdoors, surrounding the building's main body in shape of five fingers, five theme spaces with different

特色的入口广场、海博公园、滨水观景、滨水展示、绿色停车五大主题空间，各空间相互交融，构成了整体的景观。同时，室外展陈空间内展出的鸟船、大溪地号木船和752导弹护卫艇，串连了发展中的人类与海洋的依存关系。园林中来自18亿年前天津蓟州区“中上元古界地层剖面”的叠层石，更是点明了天津这座退海成陆的城市年轮和身份，让我们在记忆的沉淀里集聚前行的力量。三是展示了技术和自然之美。一百多年前，英国建筑评论家约翰·拉斯金指出：“建筑之价值所在，取决于两种截然分明的性质——其一，是建筑承继自人类，富有力量的印象；另一，则是起源于‘自然’，符合造物的形象。”（《建筑的七盏明灯》，155页，约翰·拉斯金著，谷意译，山东画报出版社，2019年出版）国家海洋博物馆总建筑面积为8万平方米，用技术之美支撑了仿自然的曲线造型。其主体结构由112榀水平与垂直方向按不同角度设置的门式钢桁架组成，高度自25米至33米不等，跨度30米左右，为布展提供了收放自如、富于张力的无柱空间，观众观展视野开阔，可以感受到技术的震撼力和冲击力。尤其是为了配合建筑向海面延伸的要求，四个展厅端部结构悬挑于海面之上，屋顶悬挑最大长度达55米。结构工程师和建筑施工团队用新技术、新材料和新工艺，解决了超长悬挑结构的施工难题，体现了建筑力与美的完美结合。为了实现博物馆非线性双曲面的外观造型，全过程采用建筑信息模型（BIM）技术，有序设计控制5.5万平方米金属幕墙的设计和安装，形成五个造型灵动的曲面屋顶。其金属

functional characteristics, namely, the entrance square, the Maritime Museum Park, the waterfront view, the waterfront display and the green parking, are formed, which blend with each other to constitute a holistic landscape. Meanwhile, a bird boat, the Tahiti wooden boat and the 752 missile escort boat displayed in the outdoor exhibition spaces present the human dependence on oceans. The stromatolites from the "Middle-Upper Proterozoic Stratigraphic Profile" of Jizhou District in Tianjin 1.8 billion years ago illustrate the annual rings and identity of Tianjin, a city developing on a land formed when the sea receded, which let us gather our strength to go forward with the profound support of accumulated memories.

Third, the museum presents the beauty of technology and nature. Over 100 years ago, John Ruskin, a British architectural critic, observed, "The value of architecture depended on two distinct characters: the one, the impression it receives from human power; the other, the image it bears of the natural creation." (*The Seven Lamps of Architecture*, p.155, written by John Ruskin, translated by Gu Yi, published by Shandong Pictorial Publishing House in 2019.) The total floor area of the museum is 80,000 m^2, with the curve modeling imitating nature supported by the beauty of technology. Its main structure is composed of 112 portal trusses arranged at different angles in horizontal and vertical directions, with a height of 25~33 meters and a span of about 30 meters, providing exhibitions with a freely retractable and tension-free space without columns, and offering visitors a wide field of vision and a feeling of amazement at technology.

Especially, in order to meet the requirements of the building extending to the sea, the end structure of four exhibition halls is overhanging the sea, with the maximum length of the overhanging roofs amounting to 55 meters. Structural engineers and the construction team have addressed the challenge of super-long overhanging structure with new technologies, materials and processes to present the perfect combination of architectural power and beauty. For realizing the appearance modeling of the non-linear hyperboloid of the museum, the Building Information Modeling(BIM) technology was adopted in the whole process to orderly control the design and installation of metal curtain walls covering an area of 55,000 m^2,

表面在不同的时间和气象下，呈现霞光流金、白灿如银的不同色相，宛如海之精灵。

“生当如鹏起，终当如鲸落”——站在国家海洋博物馆的临海平台上，仰望博物馆探向渤海的巨大悬挑结构，国家海洋博物馆里关于“鲸落”的描述浮现在我脑海中。作为地球上最大的哺乳动物，鲸鱼自由穿梭在一望无际的大海中，当它们从海中一跃而起，身姿优美，宛若鲲鹏翱翔天际。传说，当鲸预感到自己的生命即将终结时，便会悄悄寻一片宁静海域等待死亡，其庞大的身躯会慢慢坠落海底，化身为各种生物所需的养分，为大海做出最后的贡献。鲸落，正是人们赋予它生命消失再开始的一个唯美而浪漫的名字。我个人认为鲸落是鲸以特殊的方式再次重生。建筑活动亦如鹏起、鲸落。设计之初的神思飞扬、建设时期的如火如荼犹如鹏起，任凭在蓝天展翅翱翔。建筑完成交付使用犹如鲸落，人们集中智慧与力量完成创造，提供了一个新的“场所”，建筑结束了图纸使命，开始了新的使用生命，并以此满足人类生活和精神享受。国家海洋博物馆的功能空间以一种“开放形”的不规则状态，面朝大海，人们赋予她“跃向大海的鱼群”“停靠岸边的船只”“伸出的张开手掌”及“海洋的多种生物”等意象。海洋面积占地球表面积的 71%，是地球最大的生态系统，生命起源于海洋，因此我更想说，她是梦想鲸灵，来自海洋，为人类提供蓝色养分；向海而生，实现海洋强国的梦想！

forming five curved roofs with vivid modeling. Under different time and weather conditions, the metal surfaces take on different hues such as flowing gold of sunglow and silver white of moonlight, as if the sea spirit was present.

"One should be born like a roc, and pass away like the whale's fall." Standing on the coastal platform of the National Maritime Museum of China, looking up at the museum's huge overhanging structure facing the Bohai Sea, I think of the description in the museum about the "whale's fall". As the largest mammal on the earth, a whale travels freely in the boundless sea. When it jumps out of the sea, it is as beautiful as a roc (an enormous legendary bird transformed from a gigantic fish) soaring in the sky. According to legend, when a whale anticipates the end of its life, it will quietly retreat to a serene sea area to wait for its death, with its huge body slowly falling to the bottom of the sea to be nutrients for various creatures, making its final contribution to the sea. Jingluo (whale's fall) is a beautiful and romantic phrase given by people to describe the disappearance of a whale's life before its rebirth. Personally, I think that "whale's fall" is the rebirth of a whale in a special way. Construction activities resemble the birth of a roc and the decline of a whale.

Designers' sparkling ideas in the beginning of design and the bustling construction period resemble the "birth of a roc", soaring freely to the blue sky. The delivery of a building is like the "whale's fall", when people concentrate their wisdom and strength to complete the creation and provide a new "place", and the building ends its mission on the design paper to start a new life of use, thus satisfying people's needs in their material and spiritual lives. The functional spaces of the museum is in an open and irregular state facing the sea, to which people put images such as "a school of fish jumping to the sea", "ships berthing on the shore", "an outstretched open palm", "diversified creatures in the ocean", etc. Covering 71% of the earth's surface, the oceans form the largest ecosystem on the earth. Life originated from the oceans. Therefore, I would like to end my article by saying that the National Maritime Museum of China is a dreamy whale spirit from the oceans, providing blue nutrients for human beings; and born facing the sea, it calls for realizing the Chinese dream of becoming a maritime power!

目录

Contents

北冥有鱼　筑梦为鲲
——天津 · 国家海洋博物馆的设计解读

In the Northern Ocean, I Build the Dream into a Marvelous Work Design Interpretation of the National Maritime Museum of China in Tianjin

刘景樑（全国工程勘察设计大师、天津市建筑设计院名誉院长）
Liu Jingliang (National Master of Engineering Survey and Design, Honorary President of Tianjin Architectural Design Institute)

历经六年的建设，天津 · 国家海洋博物馆对外开放。国家海洋博物馆是中国首座以海洋文化为主题的国家级、综合性、公益性的博物馆，在致力于增强“中华民族海洋意识”的同时，更倡导“和谐海洋 · 和谐世界”的核心理念。它的建设体现了中国海洋发展战略，承担了重塑中国海洋价值观的重任，是集历史性、文化性、艺术性于一身，集收藏保护、展示教育、科学研究、文化传播、休闲旅游于一体的教育基地、科普基地、科学研究基地和海洋文化研究与传播中心。它不仅是天津市全新的文化地标，更是建设海洋强国，提升中国海洋文化遗产保护大境界的文化里程碑。作为该项目的设计总负责人，我见证它一步步完美地呈现，那些令人眼前一亮、耳目之新的馆内外空间、博物馆丰富展陈的文化表达，都会打开我记忆的闸门，至今仍感到振奋。据此，借本书的

On September 16, 2020, after six years of construction, the National Maritime Museum of China in Tianjin was officially opened to the public, which is the first national, comprehensive and public museum themed maritime culture in China. While being committed to enhancing the maritime awareness of the Chinese Nation, the museum devotes more efforts to advocating the core concept of "harmonious ocean and harmonious world", embodying China's maritime development strategy and undertaking the important mission of reshaping China's maritime values. It is an educational base, popular science base, research base and maritime culture research and communication center featuring historical, cultural and artistic characteristics, and integrating such functions as collection, conservation, exhibition, education, scientific research, cultural communication and recreational tourism. The museum is not only a brand-new cultural landmark in Tianjin, but also a cultural milestone in China's efforts to build a maritime power and pass on the great vision of protecting China's maritime cultural heritages. As the chief designer of the project, I have witnessed its development to perfection step by step. The amazing indoors and outdoors spaces that lighten up people's eyes, the wonderful

2016 年 5 月 11 日，国家海洋博物馆专家委员会单霁翔主任（左 6）在北京天泰宾馆主持召开第三次专家工作会议后合影
On May 11, 2016, Director Shan Jixiang (sixth from left) of the Expert Committee of the National Maritime Museum of China took a group photo after hosting the third expert working meeting at Beijing Tiantai Hotel

开篇之机归纳介绍如下。

一、筹建过程

2007 年 9 月，国内三十余位院士联名上书时任国务院总理温家宝，提议我国兴建一座国家级海洋博物馆。

2008 年，国务院批准实施的《国家海洋事业发展规划纲要》将建设国家海洋博物馆列入其中。

2010 年 4 月，天津凭借“近海临都”的区位优势和深厚的海洋文化底蕴，成为承建国家海洋博物馆入选城市。经国务院批准，国家发改委下发文件，将项目最终选址在天津滨海新区。2011 年天津市建筑设计院开始配合规划部门进行国家海洋博物馆的立项设计筹备工作。

天津市委、市政府高度重视国家海洋博物馆的建设工作，将它列为落实科学发展观、加强国家海洋文化建设、

cultural expressions of the diversified exhibitions in the museum will all open the door of my memory, and keep me exited even now. I would like to make an introduction to the museum at the beginning of this book.

I. The preparatory process

In September 2007, over 30 academicians in China jointly wrote a letter to Premier Wen Jiabao of the State Council, proposing to build a national maritime museum in China.

In 2008, the *Outline of the National Maritime Development Plan* approved by the State Council included the construction of the National Maritime Museum of China.

In April 2010, Tianjin was selected to construct the National Maritime Museum of China by virtue of its location advantage and profound maritime cultural heritages. With the approval of the State Council, the National Development and Reform Commission issued a document to finally locate the project in Tianjin Binhai New Area. In 2011, the Tianjin Architectural Design Institute (TADI) began to cooperate with the planning department to prepare for the project design of the National Maritime Museum of China.

2014 年 4 月 22 日，洛德公司专家（左起）哈维尔 · 吉梅内斯、盖尔 · 德克斯特 · 洛德（总裁）、宋汝棻、关键研讨展陈设计方案

On April 22, 2014, the experts Javier Jimenez, Gail Dexter Lord (CEO), Song Rufen, and Guan Jian (starting from the left), from Lord Company, discussed the exhibition design scheme

2014 年 9 月 14 日，国家海洋博物馆专家委员会主任单霁翔先生在北京故宫博物院主持召开第二次专家会

On September 14, 2014, Mr. Shan Jixiang, Director of the Expert Committee of the National Maritime Museum of China, presided over the second expert meeting at the Palace Museum in Beijing

2019 年 3 月 29 日，天津市副市长孙文魁一行视察国家海洋博物馆（组图）

On March 29, 2019, Sun Wenkui, Vice Mayor of Tianjin and other government officials inspected the National Maritime Museum of China (photos)

提升城市文化品位的重大工程，旨在提升全民族海洋意识，践行现代海洋观，推进我国海洋事业发展和海洋强国战略实施。市政府特别组织以市规划局为主，滨海新区政府、市海洋局配合开展园区概念性城市设计及国家海洋博物馆建筑方案的国际征集工作。

2012 年 8 月，国际征集发布会。

2012 年 9 月，中期交流会。

The Tianjin Municipal Party Committee and Municipal Government attached great importance to the construction of the museum, and listed it as a major project to implement the scientific outlook on development, strengthen the construction of national maritime culture and enhance the cultural taste of cities, aiming at boosting the maritime awareness of the whole nation, practicing the modern maritime concept, promoting the development of China's maritime industry and implementing the strategy of becoming a maritime power. The Municipal Government specially organized the

2012 年 11 月，国际征集评审会。

2012 年 12 月，国际征集方案深化。

2013 年 3 月，国际征集终审会，确定建筑方案。

终审结果是，由天津市建筑设计院（TADI）和澳大利亚 COX 建筑师事务所组成的联合团队，在与国内外 6 家设计团队的两轮比选中一举胜出。一座与我国海洋大国地位相匹配的现代化综合性国家海洋博物馆的设计方案由此诞生。当年，由天津市政府和国家海洋局联合组成国家海洋博物馆管理委员会，领导建馆的全面工作，并成立由时任故宫博物院院长的单霁翔任主任的专家委员会，以指导展陈大纲制订和审定提升展陈设计方案的工作。

二、工程概况

国家海洋博物馆地处天津市滨海新区中新天津生态城南湾的海洋文化公园内，用地面积 15 公顷，建筑面积 8 万平方米。中新天津生态城建于 2008 年，是中国和新加坡两国政府的战略性合作项目。该生态城的建设为资源节约型、环境友好型社会提供了积极探讨和典型示

Municipal Planning Bureau, and the Binhai New Area Government and the Municipal Oceanic Administration to carry out the international solicitation of conceptual urban design of the park and the architectural scheme of the National Maritime Museum of China.

August 2012, International Solicitation Release Conference.

September 2012, Mid-term Exchange Meeting.

November 2012, International Solicitation Review Meeting.

December 2012, International Solicitation Scheme Deepening.

March 2013, the construction plan determined at the International Solicitation Final Review Meeting.

The final result is that the joint team composed of the TADI and Australian COX Architect won the two rounds of competition with six design teams at home and abroad. The design scheme of a modern and comprehensive national maritime museum matching the status of China as a maritime power came into being. In the same year, the Tianjin Municipal Government and the State Oceanic Administration jointly established the National Maritime Museum Management Committee, which led the overall museum construction. Meanwhile, an expert committee headed by then President Shan Jixiang of the Palace Museum was established to guide the work of formulating the exhibition outline and examining and upgrading exhibition design schemes.

II. Project overview

The National Maritime Museum of China is located in the

设计团队现场服务（组图）
On-site service of the design team (photos)

2016 年 4 月 19 日，刘景樑大师在天津市建筑设计院和菲利普 · 考克斯教授共商海洋博物馆设计事宜
On April 19, 2016, Master Liu Jingliang discussed the design of the Maritime Museum with Professor Philip Cox at the Tianjin Architectural Design Institute

范。其中，南湾公园是生态城内重要的标志性海景主题公园，水域面积逾 280 公顷，岸线总长度约 13 千米。“海”是南湾公园的一大特色。南湾公园的水域与外海相通，随着潮汐变化，增添了海洋文化的韵味和灵动感。国家

Maritime Culture Park of Nanwan, China-Singapore Tianjin Eco-city, Binhai New Area, with a land area of 15 hectares and a floor area of 80,000 m^2. China-Singapore Tianjin Eco-city, established in 2008, is a strategic cooperation project between the governments of China and Singapore. The construction of this eco-city is an active exploration of and typical demonstration for a resource-saving and environment-friendly society. The Nanwan Park is an important landmark seascape theme park in the eco-city, with a water area of over 280 hectares and a total coastline length of about 13 kilometers. “Sea” is a major feature of Nanwan Park. The water of the park is connected with the open sea, which, with the tidal changes, enhances the charm and spirit of maritime culture. The National Maritime Museum of China is to connect the northern bay with the identity of “Green Three-Star” certification, integrate with the surrounding green environment, and become the most distinguished and expressive member to the Nanwan Park. As a matter of fact, today’s National Maritime Museum of China is unique in its multicultural expression of content and form, layout and display, concept and realm, symbol and implication, abstraction and concreteness, phenomenon and essence, commonality and individuality, presenting the uniqueness and innovation of the immense epoch-making Chinese maritime culture.

2016 年 9 月 23 日，国家海洋博物馆筹建办公室接待大英博物馆副馆长加雷斯 · 威廉姆斯（左 8）所率团队考察国家海洋博物馆工程现场
On September 23, 2016, the Preparatory Office of the National Maritime Museum of China received Gareth Williams (the eighth from left), Deputy Director of the British Museum, and his team to visit the project site of the museum

建设中的国家海洋博物馆
The National Maritime Museum of China under construction

海洋博物馆正是要以“绿色三星”认证的身份与北部海湾相交汇，与周边绿化环境相融合，为南湾公园增添最尊贵且具有丰富表现力的新的一员。事实上，今天的国家海洋博物馆确在内容与形式、布局与展示、理念与境界、象征与寓意、抽象与具体、现象与本质、共性与个性的多元文化表达上独树一帜，呈现了丰富中国海洋文化所展示出的时代性、特色性与创新性。

三、设计特色

“馆园融合”是国家海洋博物馆规划设计布局的总思路。大境界之所以可铸就高品位，重在要求建筑师要站在历史与时代的高度，在用深邃的目光去品读海博馆博大精深定位的同时，拥有创作的品质与境界。这确非

III. Design features

"Museum-park integration" is the general idea for the layout of the National Maritime Museum of China. A great vision can create a great work of great taste because architects are required to stand at the height of history and the times, and to read the profound positioning of the museum with great insights. This is indeed not a ordinary cultural project. From the perspective of the national ocean conception, it is a solid work that clearly defines China's maritime cultural heritages, offering a big stage for the public to understand the long history of Chinese maritime culture and the harmonious coexistence of contemporary maritime strategies to the industry and the Chinese and foreign visitors; in light of the blueprint of Tianjin's urban culture development, the scientific design of the museum can fully embody the vision of developing "Cultural Tianjin", telling the world that the "Tianjin expression" and "city spirit" enlightening the mind can be found here.

Therefore, we should encourage the design and creation of a contemporary classic of maritime architecture. The design will

国家海洋博物馆入口广场
Entrance square of the National Maritime Museum of China

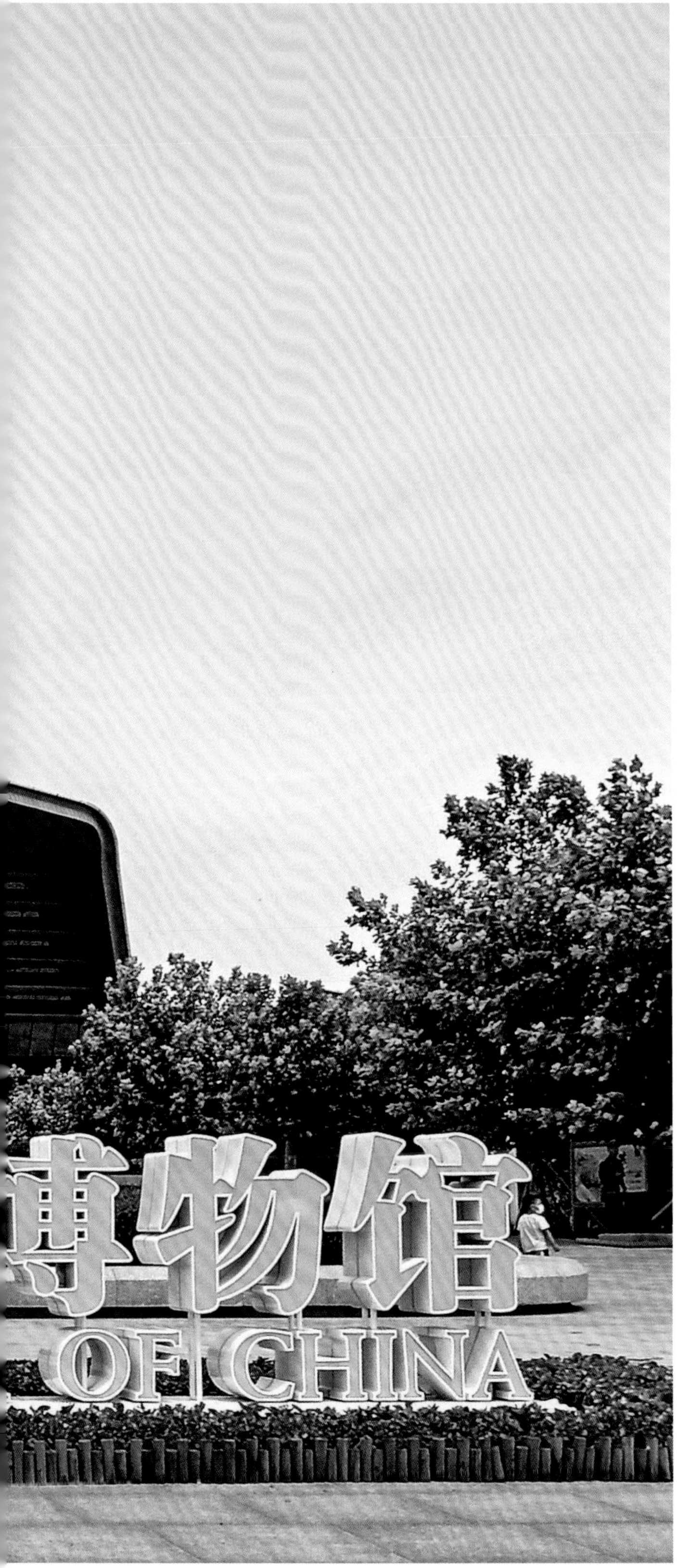

一般的文博项目，从国家海洋观来看，它是清晰界定中国海洋文化遗产内涵的厚实之作，它是向业界与中外公众普及中国海洋文化悠远历史脉络与当代海洋战略和谐共生观的大舞台；从天津城市文化发展蓝图上看，国家海洋博物馆的科学设计恰恰可更充分地体现“文化天津”建设之愿景，它将告知世人，在这里可领略到启迪心智的“天津表情”与“城市精神”。据此，要激励设计创造追求海洋建筑的当代经典。设计将建筑与景观和场所整体营造，着力体现在以下三个方面：一是，用建筑架构陆路与水路的空间平台交织；二是，实现建筑与五大空间无缝对接的完美融合； 三是，外展与内展场所的一体化统筹，它也符合单霁翔院长的“从馆舍天地走向大千世界”的广义博物馆之思。“馆园融合”的设计思路彰显博物馆区成为整合中新天津生态城新城区、南湾生态文化公园和海洋自然水域的一个空间界面，并成为人类与海洋互动的新场所、新空间的媒介。

“开放形”是博物馆个性化建筑的形态特质，是设计

create architecture, landscape and place as a whole, which is reflected in the following three aspects. First, the spatial platform of land and waterway are intertwined; second, the perfect integration and seamless connection are realized between the building and the five major spaces; third, the integration and overall planning of outdoors and indoors exhibition venues are realized, which is also in line with President Shan Jixiang's idea of a museum in a broad sense, namely "breaking through the confines of a museum building to the wide world". The design idea of "museum-park integration" is reflected in the fact that the museum area has become a spatial interface integrating the new urban area of China-Singapore Tianjin Eco-city, Nanwan Eco-cultural Park and the natural sea waters, turning into a new space medium for interaction between human beings and the sea.

"Openness", as a morphological characteristic of personalized architecture of the museum, is the core idea of its design creativity. With the metaphorical non-linear technique, flowing architectural vocabulary, and the building unfolding from the land to the sea

中央大厅实景
Central hall

外景鸟瞰
Aerial view of the location

公共空间实景
Public space

今日海洋厅
“Today’s Ocean” exhibition hall

创意的核心理念。设计运用隐喻的非线性手法，流动的建筑语汇，建筑由陆地向海洋展开并架空在海面上，以表达三个层面的创意：一是，向海亲海，用建筑架构陆地与海洋空间的交织互动，营造对海洋的多种感受和体验，展示陆地文明对海洋文明的探索与追求；二是，融合统一，以中央大厅为基准空间，塑造发散形的建筑形态，四条侧翼展示建筑空间与多元展陈主题艺术空间环境的和谐共融，是馆内展览空间交通动线的外在体现，是建筑功能与形式的完美融合，同时也彰显结构空间力学与建筑美学的高度统一；三是，挑战自我，通过陆地文明向海洋文明探索追求的进程，展示“建筑人”不断挑战自我，以不负韶华的历史责任感，攻克多项技术难题，见证展示陆海文明载体落成的全过程。国家海洋博物馆之所以气势恢弘又不失文博建筑的典雅与文化性，贵在设计者从国家海洋文化遗产的保护、传承与创新上，下功夫深耕“作业”，精准把握住海岸遗产，包括滨海海岸独有的自然景观及人文景观等。中国海洋文化遗产

and overhanging above the sea, the designer attempts to express creativity in three aspects. First, the building faces to the sea and is close to the sea, to present an architectural structure interweaving and interacting with the sea space, create a variety of feelings and experiences of the sea, and show the exploration and pursuit of a land civilization to a maritime civilization. Second, the building features integration and unification. With the central hall as the reference space, it creates a divergent architectural form with four flanks highlighting the harmonious integration of architectural space and multi-exhibition thematic art space and environment. This is the external embodiment of the traffic dynamic line of exhibition space in the museum and the perfect integration of architectural function and form, while stressing the great unity of structural space mechanics and architectural aesthetics. Third, the building expresses self-challenge, as the process in which a land civilization explores and pursues a maritime civilization shows that "people engaged in architecture" constantly challenge themselves to overcome many technical problems and witness the whole process of completing the carrier of land and maritime civilizations. The National Maritime Museum of China is magnificent without losing the elegance and cultural attributes of a museum building. That thanks to the designers' efforts in the protection, inheritance and innovation of the national maritime cultural heritages, in an attempt to accurately grasp the coastal heritages, including the unique natural and

的主要存在空间是环中国海，强化国家海洋历史与文化认同，就是要着力于人为建设工程（含文博纪念类项目）。宣传中国是海洋强国，从而促进国民建设，提升国民历史自豪感与文化自信，使其重视海洋文化遗产传承与发展，对繁荣当代海洋文化意义重大。国家海洋博物馆的设计营造恰恰体现了历史人文及科技创新的海洋文化之根。

四、工程获奖

国家海洋博物馆是一个国际化、多样化的建筑创作成功案例。有人会问，博物馆还能做什么？通过查阅文献得

cultural landscapes of the coastal area of Binhai New Area. China's maritime cultural heritages are mainly found in seas around the China. To strengthen the national maritime history and cultural identity relies on built projects (including cultural and memorial projects) to demonstrate that China is a maritime power, so as to promote national construction, enhance historical pride and cultural self-improvement and self-confidence and make them attach importance to the inheritance and development of maritime cultural heritages. That is of great significance to the prosperity of contemporary maritime culture. The design and construction of the National Maritime Museum of China precisely reflects the roots of maritime culture which is both historical and scientific.

IV. Honors

The National Maritime Museum of China is a successful case of international and diversified architectural creation. Some people

3 号厅实景（施工中）
Hall 3 (under construction)

知，1974 年“为社会及其发展服务”的宗旨被写入国际博物馆协会大会修订章程中的博物馆新定义后，博物馆逐步告别“精英至上”的理念，成为更加亲民的社会发展与变革的服务者、参与者和推动者。同样更要求建筑师的文博空间营造，不仅更便于丰富产品的展示及文化艺术历史脉络的梳理，而且能使观者与研学者共同开启一段“补全可视世界”的美妙之旅。国家海洋博物馆的设计者以责任心和行动力做到了这一切，既体现了海洋生命之光，也表现了人文与科技之美；用设计深造进入馆舍天地，拥有高远意境的同时寓教于乐，吸引着越来越多的中外宾客，尤其提升了国民对海洋文博的兴趣。

国家海洋博物馆因设计与其所处的场地环境保持了良好的互动关系，在全球范围获得了良好的声誉和评价。在 2013 年新加坡世界建筑节（WAF）上， 该设计是唯一同时荣获三项大奖的项目，即：“最佳未来建筑奖”“最佳文化建筑奖”“最佳竞赛建筑奖”。组委会在评语中写道：“该项目展示了一个强大清晰的理念，在与海洋的互动中，唤起了游客强烈的海洋体验，承载了中国悠久的海洋历史，为游客提供了丰富的全球海事文化，评审团期待这个独特的创意能够完整地实现。”2016 年 9 月，大英博物馆副馆长加雷斯 · 威廉姆斯率英国国家海事博物馆领导、专家等一行专程来滨海新区参观，在博物馆施工现场看到当今世界规模最大的一座海洋博物馆时，他兴奋地表示：“你们的海洋博物馆不仅属于天津、属于中国，而且是属于世界的。”

国家海洋博物馆的造型是灵动而富有感染力的，在建筑设计以及与施工配合的技术要求方面都充满着挑战。传统平面二维图纸无法准确表达非线性双曲面建筑的全部空间信息，为此设计者采用 BIM 技术进行参数化计算机建模设计，成功解决了大尺度和小细节共存的设计

may ask: what else can a museum do? A literature review shows that after the purpose of "serving society and its development" was written into the new definition of museums in the revised charters of the International Council of Museums in 1974, museums have gradually broken away from the idea of "elite first" to become a more people-friendly service provider, participant and driver of social development and change. Meanwhile, further requirements are asked for architects to create a museum space which not only is more convenient for the display of diversified products and the review of the historical context of culture and art, but also enables visitors and researchers to start a wonderful journey of "complementing the visual world". The designers of the National Maritime Museum of China have accomplished all these requirements with their sense of responsibility and actions, creating a museum which reflects the light of maritime life and presents the beauty of culture, as well as science and technology. Design is further developed to create a world in the museum building and make education a joyful matter with a lofty vision, attracting more and more Chinese and foreign visitors, and in particular, enhancing Chinese people's interest in maritime cultural heritages.

The National Maritime Museum of China has gained a good reputation in the world, winning favorable comments since it has maintained a good interactive relationship with its site environment. At the 2013 Singapore World Architecture Festival (WAF), this design was the only project winning three awards at the same time, namely, "Best Future Architecture Award", "Best Cultural Architecture Award" and "Best Competition Architecture Award". The comments of the Organizing Committee go, "The project presents a strong and clear concept. In the interaction with the sea, the building evokes a strong maritime experience for tourists, carries the long history of China's maritime culture, and provides tourists with rich global maritime cultural experience. The jury expects this unique and creative idea to be fully materialized." In September 2016, Gareth Williams, Deputy Director of the British Museum, led some leaders and experts of the British National Maritime Museum to visit the Binhai New Area. At the sight of the largest maritime museum in the world under construction, he remarked excitedly, "Your maritime museum belongs not only to Tianjin, to China, but also to the world."

The shape of the National Maritime Museum of China was designed to be vividly appealing, which imposed challenges in terms of architectural design and technical requirements for cooperation with construction. As traditional two-dimensional drawings could not accurately express all the information of non-linear hyperboloid architectural space, the BIM technology was adopted to design

难题，圆满完成了富于个性化和感染力的国家海洋博物馆设计。该 BIM 设计成果获得了国内外多项设计大奖，2019 年 11 月，在拉斯维加斯举行的“全球工程建设行业卓越奖”的颁奖大会上，该项目荣获中型项目组的“最佳实践奖”第一名。

国家海洋博物馆在设计营造中获得的一系列盛誉都离不开各界的鼎力支持：当年国家海洋局宣教司主持并组织展陈大纲和展陈内容的编制工作，为建筑设计提供了奠基性与指导性的专业支持；鉴于国家海洋博物馆建设以零藏品艰难起步的复杂性，在建筑设计之初，引进了博物馆专业策划团队——加拿大洛德文化资源规划和管理公司与北京 LORD 国际文化发展有限公司，其以功能与空间充分融合的展陈策划和概念设计，与建筑设计密切对接配合，实现了有针对性的设计目标；在博物馆筹建办的领导下，金大陆文化产业（集团）有限公司、北京清尚建筑装饰工程有限公司、天禹文化集团有限公司三家设计团队，出色地完成了全部展陈规划、设计和制作；尤其感谢以单霁翔院长为主任的专家委员会，对博物馆的展陈设计组织召开多次专家委员会评审，确保了创新且多元化的高水平展陈设计方案的落实，给观者带来富于亲和力与感召力的非凡感受。

如果说，为多元化展陈主题的展览空间营造收放自如的艺术展域，是本博物馆建筑设计的创新点，那么，结构工程师们的设计，确值得大书一笔。即：结构设计通过突破诸多技术难题，对大跨度的无柱空间、跨层跃层的超高空间、超大尺度的悬挑空间等巧妙且有机整合，充分满足各展厅的多元化空间布展并提升以“公众为本”的观展需求，同时彰显出“海洋自然”和“海洋人文”两大主题展览的场域整体感和连续性的空间魅力，为公众观展提供了开阔的空间视觉享受和场景的冲击力、震

parametric computer modeling, which successfully solved the design challenge of coexistence of large-scale dimensions and small details. Thus, the design of a uniquely appealing museum was successfully completed. Meanwhile, the BIM design achievements have won many design awards at home and abroad. In November 2019, at the awarding ceremony of "Global Construction Industry Excellence Award" in Las Vegas, the project won the first place in the "Best Practice Award" of the medium-sized projects.

The honors won for the design and construction of the National Maritime Museum of China cannot be separated from the vigorous supports from all walks of life. Earlier, the Department of Communications and Education of the State Oceanic Administration presided over and organized the preparation of the exhibition outline and exhibition contents, offering foundation-laying and guiding professional support for architectural design. Given the situation that the museum had no exhibits in the first place, at the beginning of the architectural design development, the professional planning team of the museum, LORD Cultural Resources Planning and Management Company of Canada, and LORD International Cultural Development Co., Ltd. of Beijing were introduced. The exhibition planning and conceptual design with full integration of function and space were closely matched with the architectural design to achieve the design goals. Under the leadership of the Museum Preparatory Office, three design teams, namely, Jindalu Cultural Industry (Group) Co., Ltd., Beijing Qingshang Architectural Decoration Engineering Co., Ltd. and Tianyu Culture Group Co., Ltd., successfully completed all the exhibition planning, design and production. Special thanks should be given to the expert committee headed by President Shan Jixiang for organizing and holding many expert committee reviews on the exhibition design of the museum, which have ensured the implementation of innovative and diversified high-level exhibition design schemes, leaving visitors an extraordinary impression of affinity and inspiration.

If creating flexible art exhibition areas for the diversified exhibition themes is an innovation of the architectural design of this museum, the design of the structural engineers indeed deserves our special mention. With breakthroughs made in many technical problems, the structural design has skillfully and organically integrated the large-span pillarless space, the super-high-altitude space across levels, and the super-large-scale overhanging space, which fully meets the diversified spatial layout of each exhibition hall and enhances the "public-oriented" exhibition demand. Meanwhile, the design highlights the sense of wholeness and the continuous spatial charm of

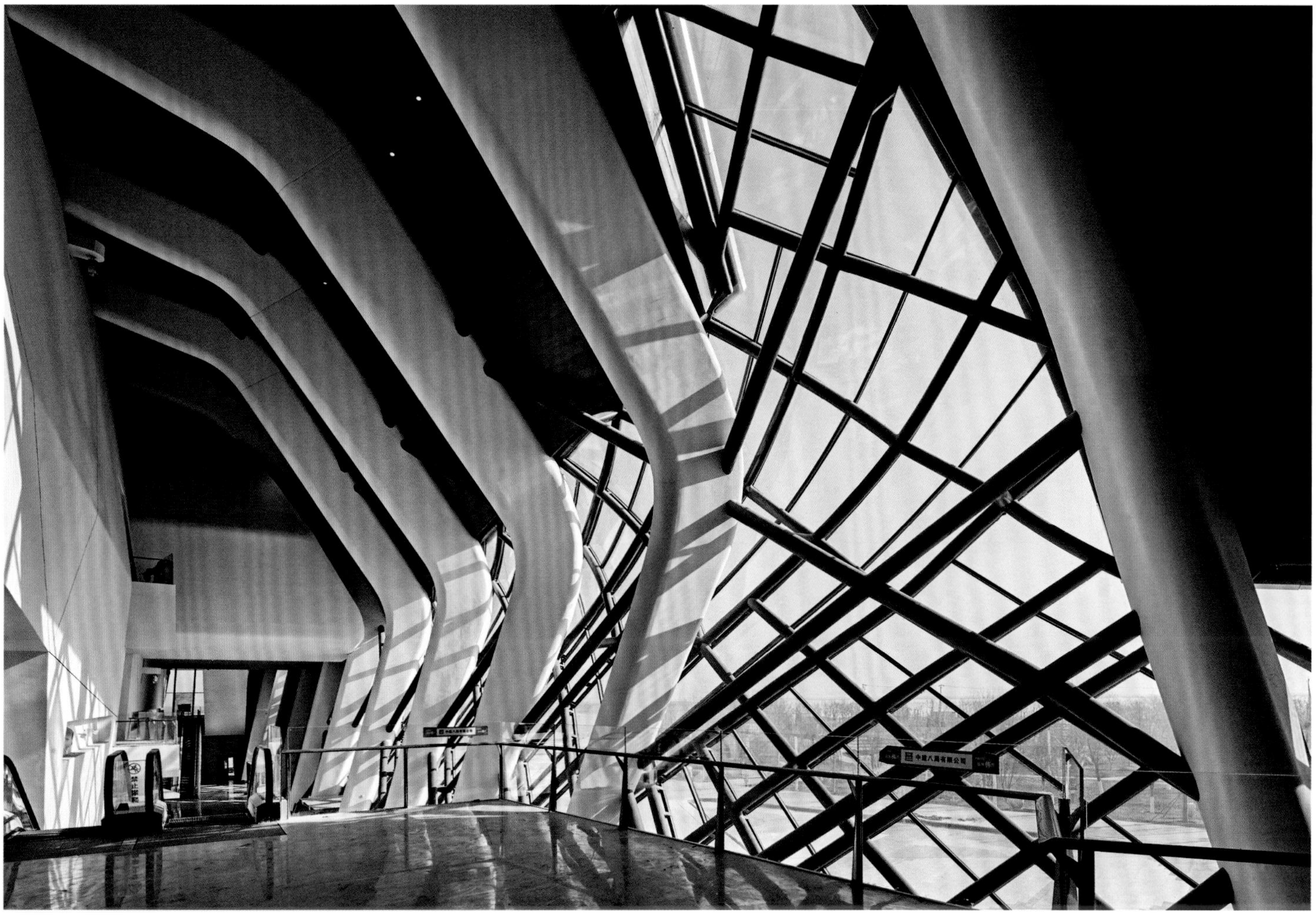

2 号厅实景
Hall 2

撼力。该项设计已分别荣获全国及天津市钢结构设计金奖。

五、筑梦为鲲

如果说概念设计是“筑梦”的起点，那么深化设计就是将梦想转化为现实的“蜕变”。在这一蜕变中，设计团队以梦为马，以汗为泉，精心设计，砥砺前行。先后攻破了“超大空间防火性能化设计”“BIM 导向性设计”“非线性表皮构造设计”“大跨度无柱空间设计”“超长悬挑结构设计”“夜景照明系统设计”等多项技术难

the two major theme exhibitions, “Maritime Nature” and “Maritime Culture”, providing expansive spatial and visual enjoyment and the impact of scenes to visitors. This design has won national and Tianjin municipal steel structure design gold awards.

V. Design development

If conceptual design is the starting point of “dream building”, design development is transforming the dream into reality. For this transformation, the design team devoted themselves to the careful design, making breakthroughs in many technical problems such as “fireproofing performance design of super-large space”, “BIM-oriented design”, “non-linear surface structure design”, “long-span pillarless space design”, “super-long overhanging structure design” and “night lighting system design”, witnessing the process of fulfilling dreams,

3 号厅实景
Hall 3

题，见证了让梦想照进现实的过程，实现了对“筑梦为鲲”的完美诠释。

如果说文化是一座城市最好的底色，那么，有创意的文博建筑一定是铺展美好城市蓝图的点睛之笔。国家海洋博物馆以“从馆舍天地到大千世界”的创意筑城，从里到外，从静态到动态的呈现，彰显了在国际视野下，国家“海洋文化全景图”的中国设计。国家海洋博物馆建成后将成为中华海洋文化标志性设施，成为国际一流的大型综合性海洋博物馆，拥有六百年历史的天津市终于迎来第一座国家级的综合性博物馆。其将极大提升天津在国内外城市中的文化品位和文化实力，靠建筑人与文博人共同的文化优势，提高国际知名度，从而引发文化、旅游、商贸、交通及相关服务业的联动发展，并成为天津市及滨海新区文化升级跨越的又一强劲引擎。虽然，人们常说城市文化地标以生活与实践积淀作为尺度，但真正体现建筑与城市之美拓新精神的文博项目，不仅会无愧于时代，更能迅速刷新人们对文化地标的观感。

谨以此书致敬向国务院联名上书的两院院士，致敬在项目设计中给予指导和关心支持的各位专家，感谢澳大利亚 COX 建筑师事务所的精诚合作，感谢天津市规划和自然资源局、原天津市海洋局的指导与支持。我相信，历史将铭记所有为国家海洋博物馆完美落成付出智慧和辛勤劳动的设计者和建设者。

2020 年 8 月

and perfectly interpreting the concept of "zhu meng wei kun"(turning a dream design into a marvelous work of architecture).

If culture is the best impression of a city, creative cultural and museum buildings must be finishing touches to present the blueprint of a beautiful city. The National Maritime Museum of China is famous for the creative idea of "breaking through the confines of a museum building to the wide world", demonstrating the Chinese design of the national "panorama of maritime culture" from an international perspective with both indoor and outdoor as well as static and dynamic presentations. After its completion, the museum will become a landmark facility housing Chinese maritime culture and a world-class large-scale comprehensive maritime museum. Tianjin, with a history of 600 years, finally witnessed the construction of the first national comprehensive museum in China. The museum will greatly enhance the Tianjin city's cultural taste and power among cities at home and abroad, and boost its international popularity by relying on the common cultural advantages of builders and scholars, thus triggering the coordinated development of culture, tourism, commerce, transportation and related service industries, and becoming another powerful engine for the cultural upgrading of Tianjin and Binhai New Area. Despite the common saying that urban cultural landmarks are measured by the accumulation of life and practice, cultural projects that truly embody the beauty of architecture and city will not only live up to the wonders of the times, but also quickly refresh people's perception of cultural landmarks.

I would like to dedicate this book to the academicians who have jointly signed a letter to the State Council, and to the experts who have given guidance, care and support during project design. I would also like to thank COX Architects of Australia for their sincere cooperation, thank Tianjin Municipal Bureau of Planing and Natural Resources for their guidance and support. I believe that history will remember all the designers and builders who have contributed their wisdom and hard work to the perfect completion of the National Maritime Museum of China.

August 2020

第一章　方案形成

Chapter I
Scheme Development

为高品质建设国家海洋博物馆，该工程项目的“概念性城市设计”及“建筑方案设计”国际征集工作于 2012 年 8 月展开。经专家组两轮评比遴选，2013 年 3 月，由澳大利亚 COX 建筑师事务所和天津市建筑设计院（TADI）组成的联合团队一举胜出。在联合团队的努力下，方案经不断完善并深化，体现了形体关系、功能布局、参观流线的三个优化，以及与使用需求、专业规范、展陈策划的三个对接，从而完满实现设计方案的高水平提升。

The international solicitation for "conceptual urban design" and "architectural scheme design" of the project started in August 2012, for the high-quality construction of the National Maritime Museum of China. In March 2013, the joint team consisting of Australian COX Architects and Tianjin Architectural Design Institute won the bid after two rounds of evaluation by the expert group. Thanks to the efforts of this team, the scheme has been constantly improved and developed, embodying the optimization of three aspects, namely physical relationship, functional layout and visiting streamline, and the integration of three aspects, namely using demans, professional standards and exhibition planning, thus perfectly upgrading the design scheme to a high level.

为了高水平完成国家海洋博物馆的建设工作，天津市规划和自然资源局滨海新区分局、国家海洋博物馆筹建办自 2012 年 8 月起，对国家海洋博物馆工程项目组织了“园区概念性城市设计”和“建筑方案设计”的国际征集工作。

征集工作分两个层面：① 概念性城市设计；② 建筑方案设计。

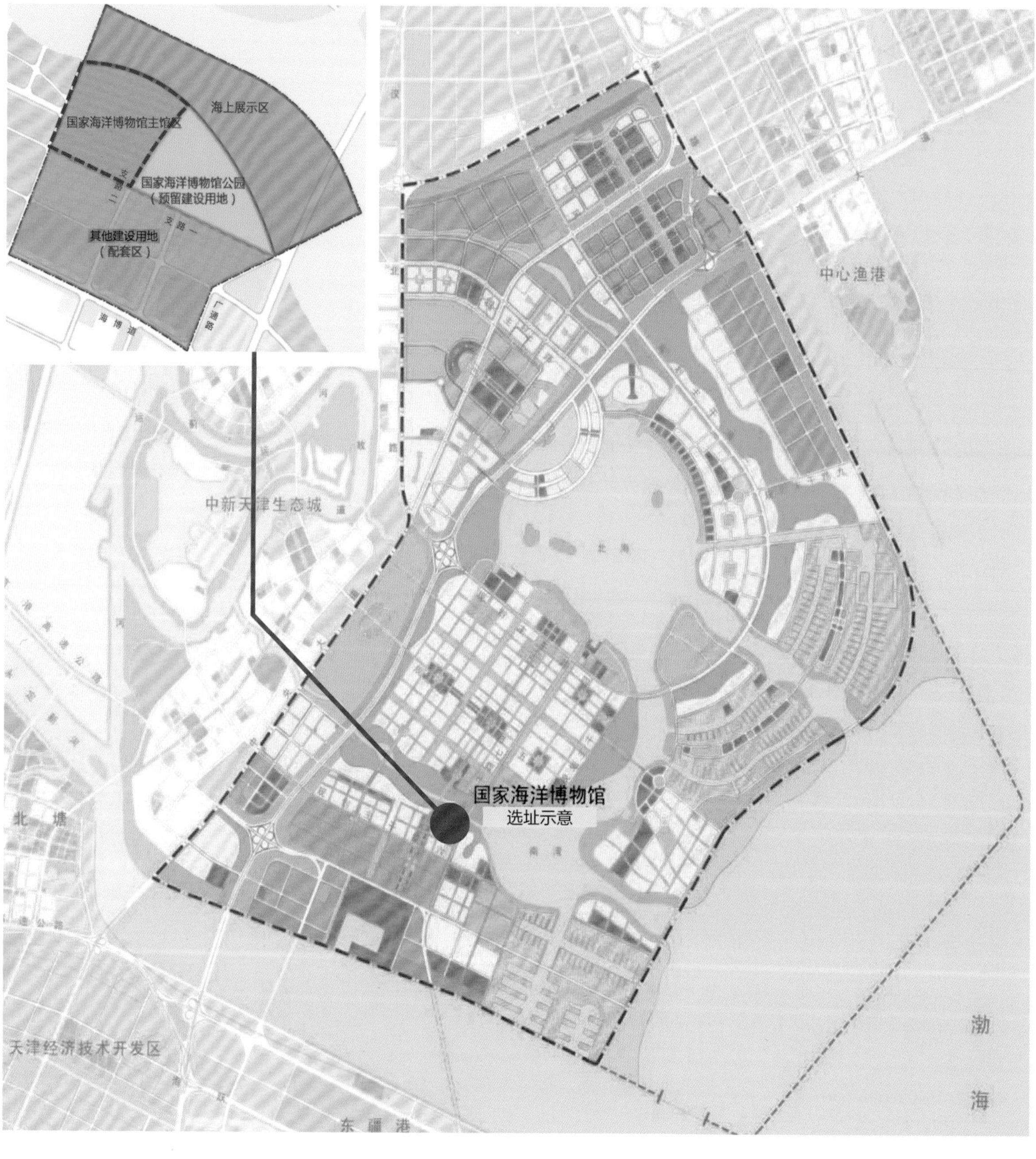

项目位于天津市滨海新区中新天津生态城南湾，在规划面积为 1 平方千米的南湾公园内

此次征集活动邀请了来自澳大利亚 COX 建筑师事务所、中国华南理工大学建筑设计研究院、西班牙 EMBT 建筑师事务所、英国沃特曼国际工程公司、德国 GMP 国际建筑设计有限公司、美国普雷斯顿·斯科特·科恩设计公司等的国内外著名设计团队参与设计。

2012 年 11 月召开概念性城市设计方案评审会，邀请崔愷院士等 10 位专家组成评审委员会

2013 年 2 月 28 日召开建筑设计深化方案评审会，邀请马国馨院士等 10 位专家组成评审委员会

澳大利亚 COX 建筑师事务所方案

中国华南理工大学建筑设计研究院方案

西班牙 EMBT 建筑事务所方案

英国沃特曼国际工程公司方案

概念性城市设计方案经专家评审投票后，选出四个入围方案，来自以下团队：澳大利亚 COX 建筑师事务所、中国华南理工大学建筑设计研究院、西班牙 EMBT 建筑师事务所、英国沃特曼国际工程公司。

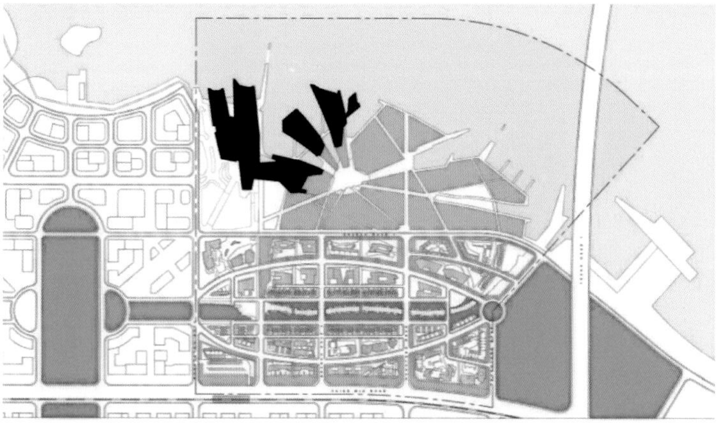

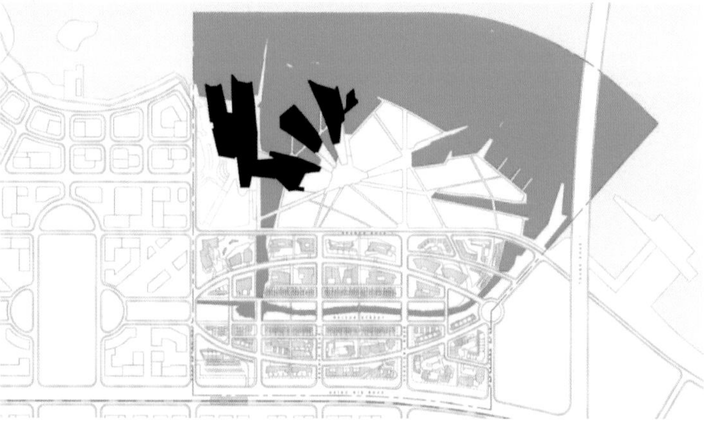

概念性城市设计方案评选阶段的专家评语：“澳大利亚 COX 建筑师事务所的设计方案突出博物馆的“海洋”主题，众多形似桥梁连接的博物馆展馆能够营造出对水的多种不同感受。设计意在建立一系列独特、相连的展馆，他们的设计让人联想到有码头和防波堤的海港、有船只停靠的安全海港、跃向海湾的鱼群，任访客自由诠释。”

俯瞰国家海洋博物馆

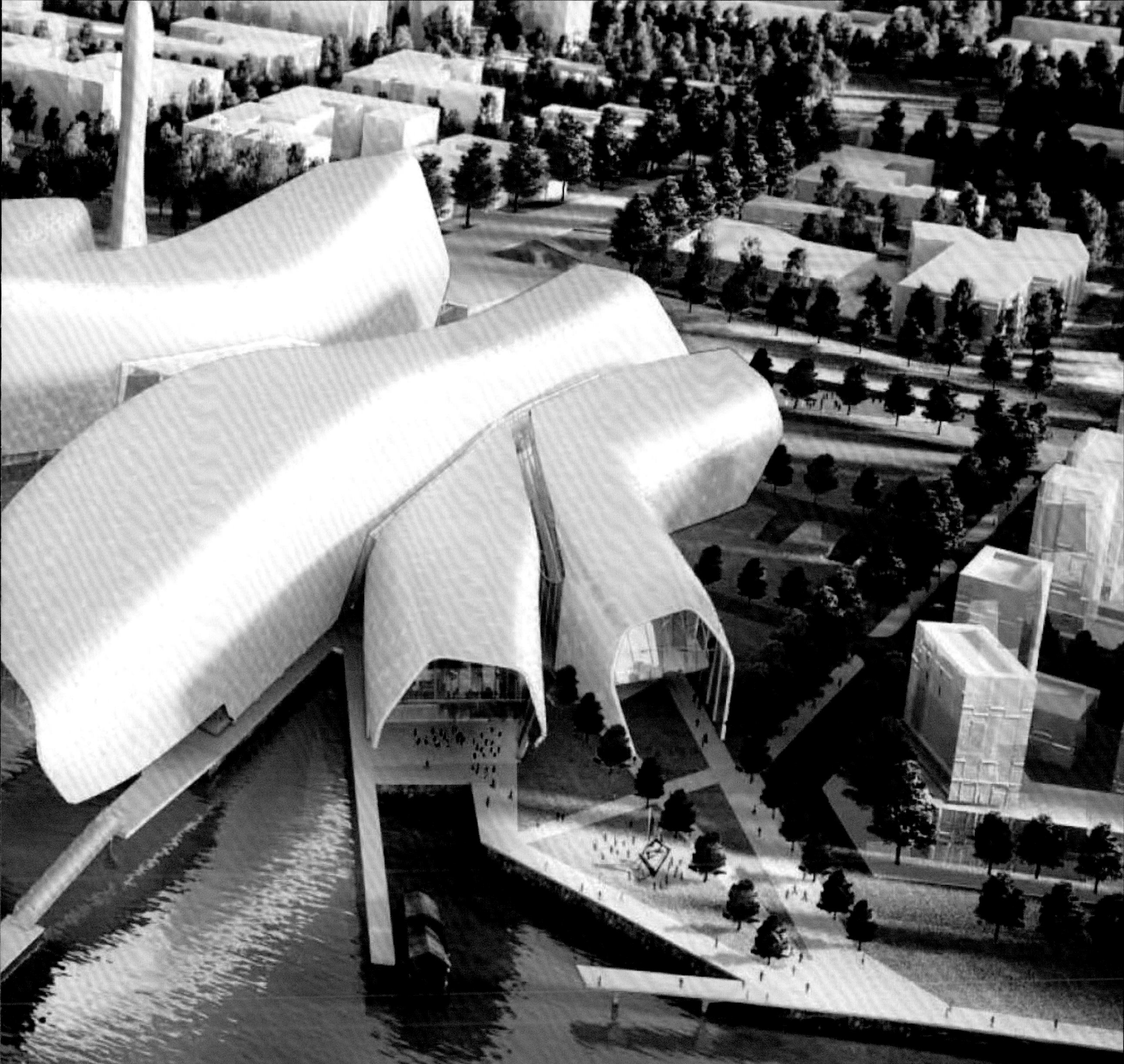

在 2012 年 11 月后的深化设计阶段，澳大利亚 COX 建筑师事务所与天津市建筑设计院（TADI）联合体结合专家意见，根据政府部门要求和建设单位需求，进行了设计的三个优化和三个对接，即：优化形体关系、优化功能布局、优化参观流线以及对接专业规范、对接展陈策划、对接使用需求，以优化和深化的设计提升，满足可实施性。深化后的方案于 2013 年 9 月通过了政府部门的审查和批准。

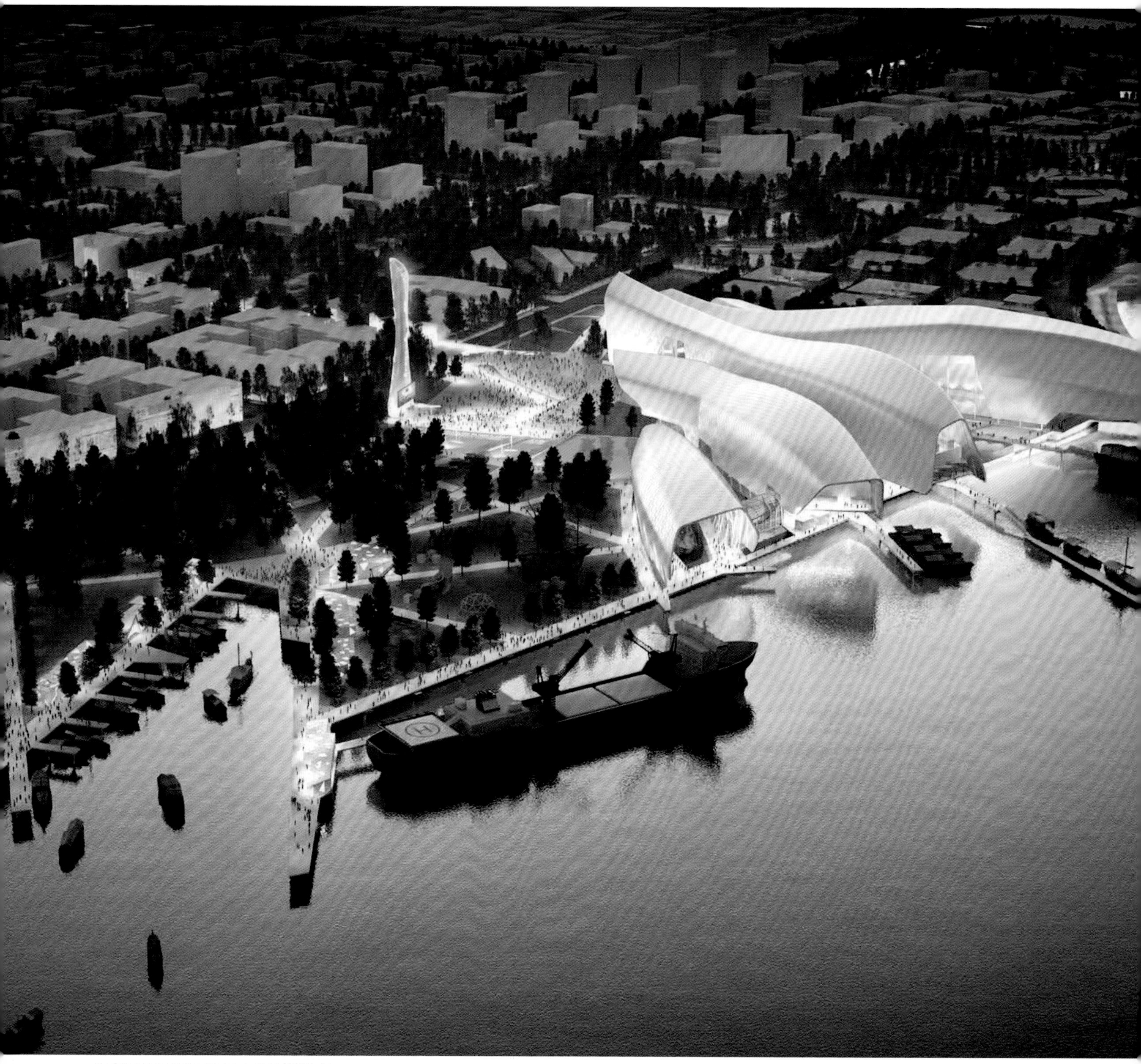

方案深化过程中的国家海洋博物馆

中国国家海洋博物馆
NATIONAL MARITIME MUSEUM OF CHINA

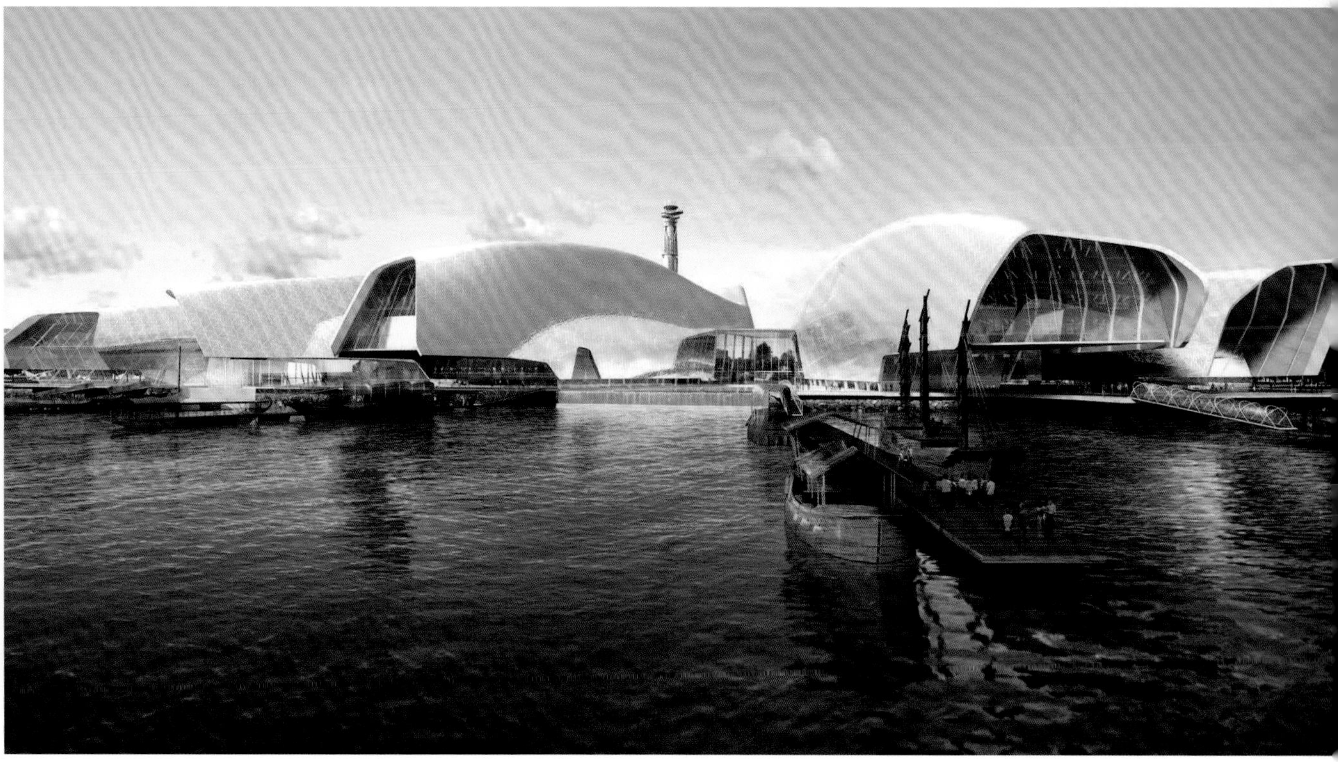

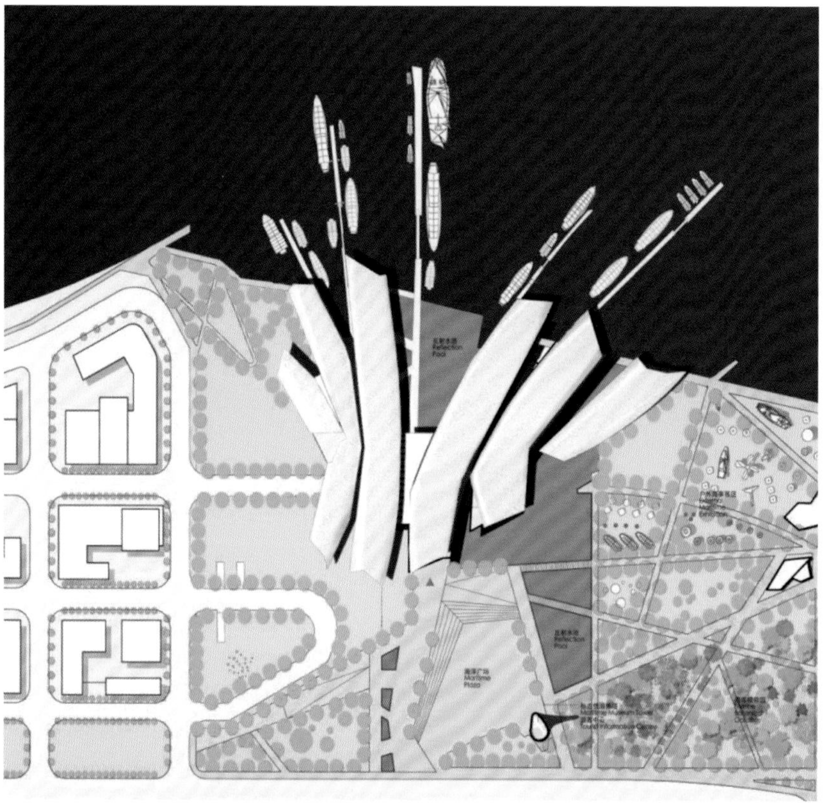

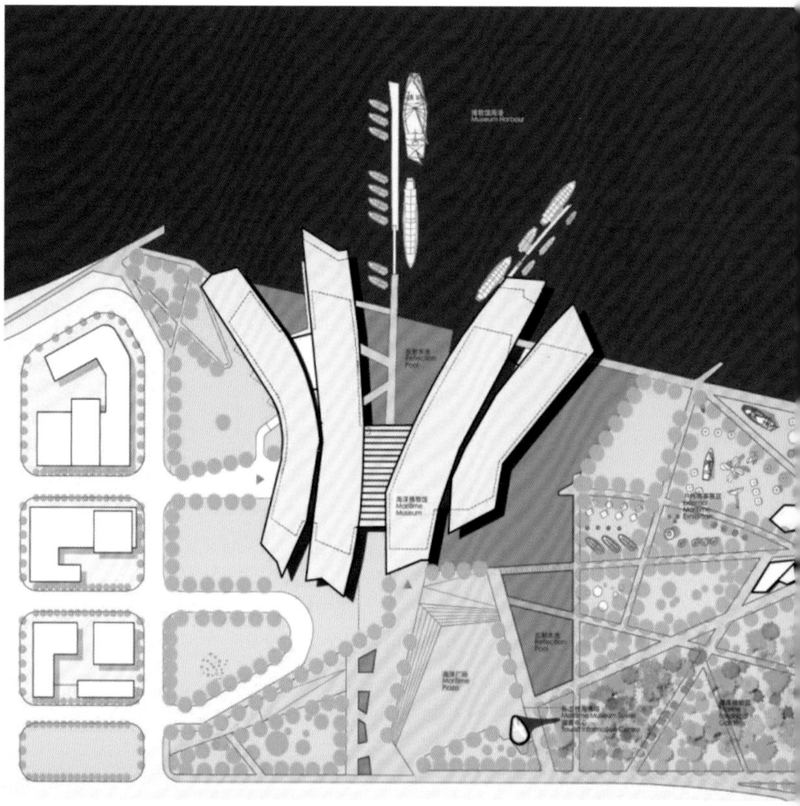

2013 年 2 月方案及效果图

2013 年 3 月方案及效果

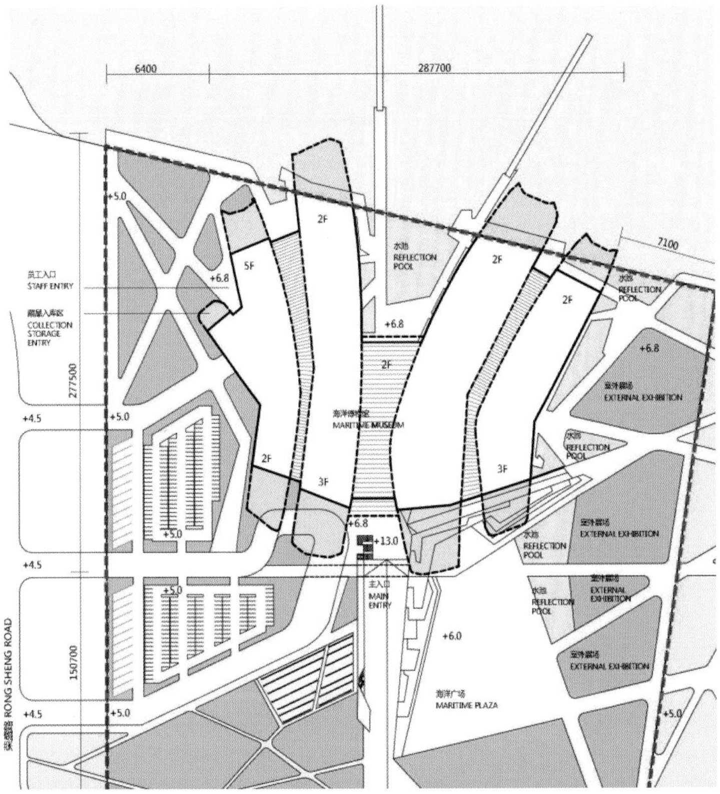

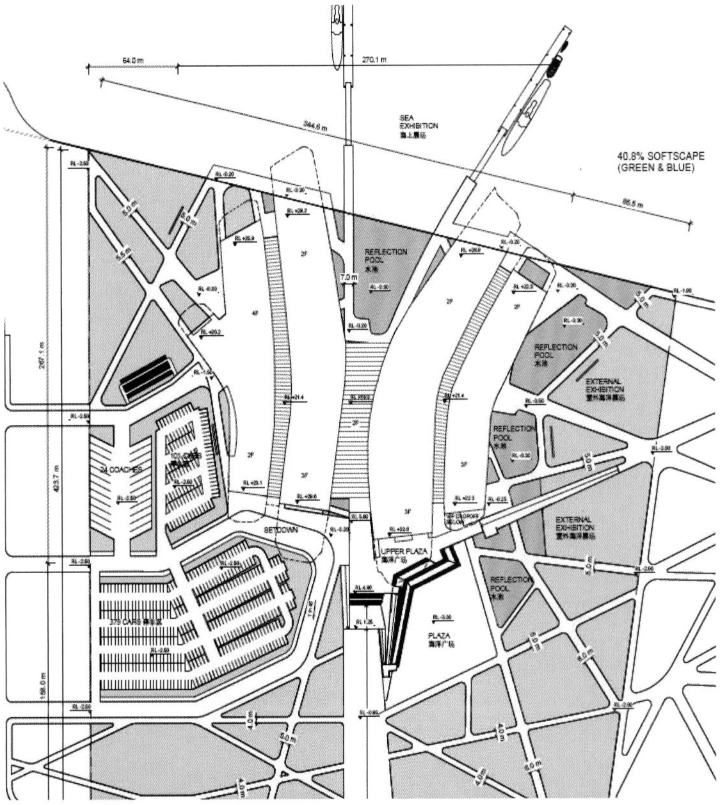

2013 年 7 月方案及效果图

2013 年 11 月方案及效果图

深化过程中的设计方案

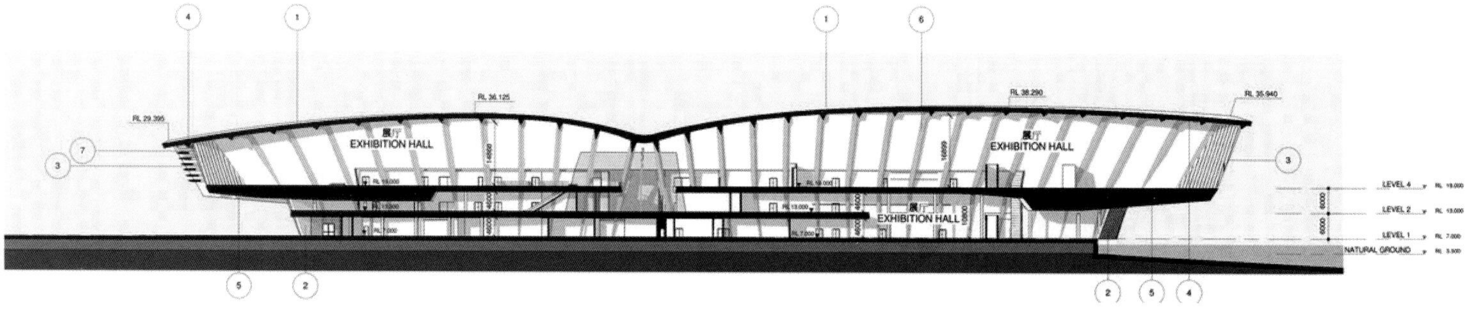
展厅
EXHIBITION HALL
展厅
EXHIBITION HALL
展厅
EXHIBITION HALL
NATURAL GROUND

报告厅
3D CINEMA
展厅
EXHIBITION
HALL
展厅
EXHIBITION HALL
NATURAL GROUND

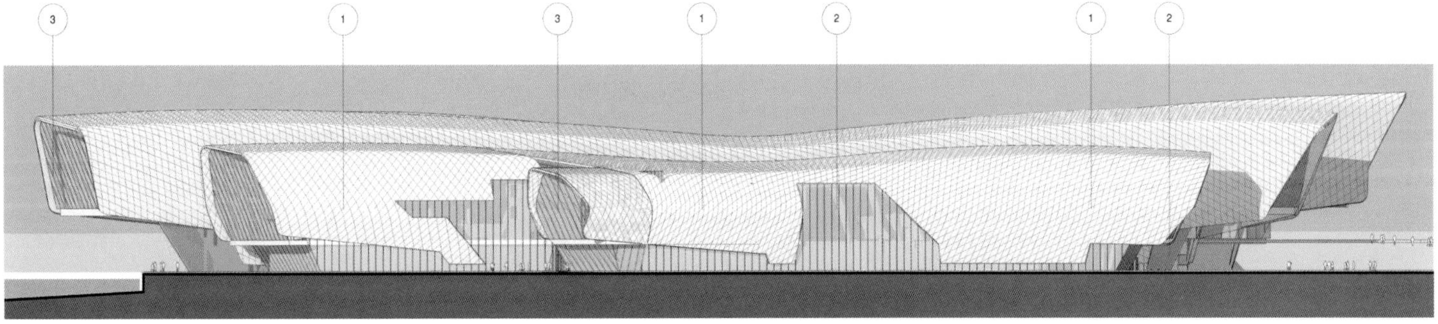
3
1
3
1
2
1
2

1
3
2
7
8
7
1
7
2

国家海洋博物馆设计方案效果图 1

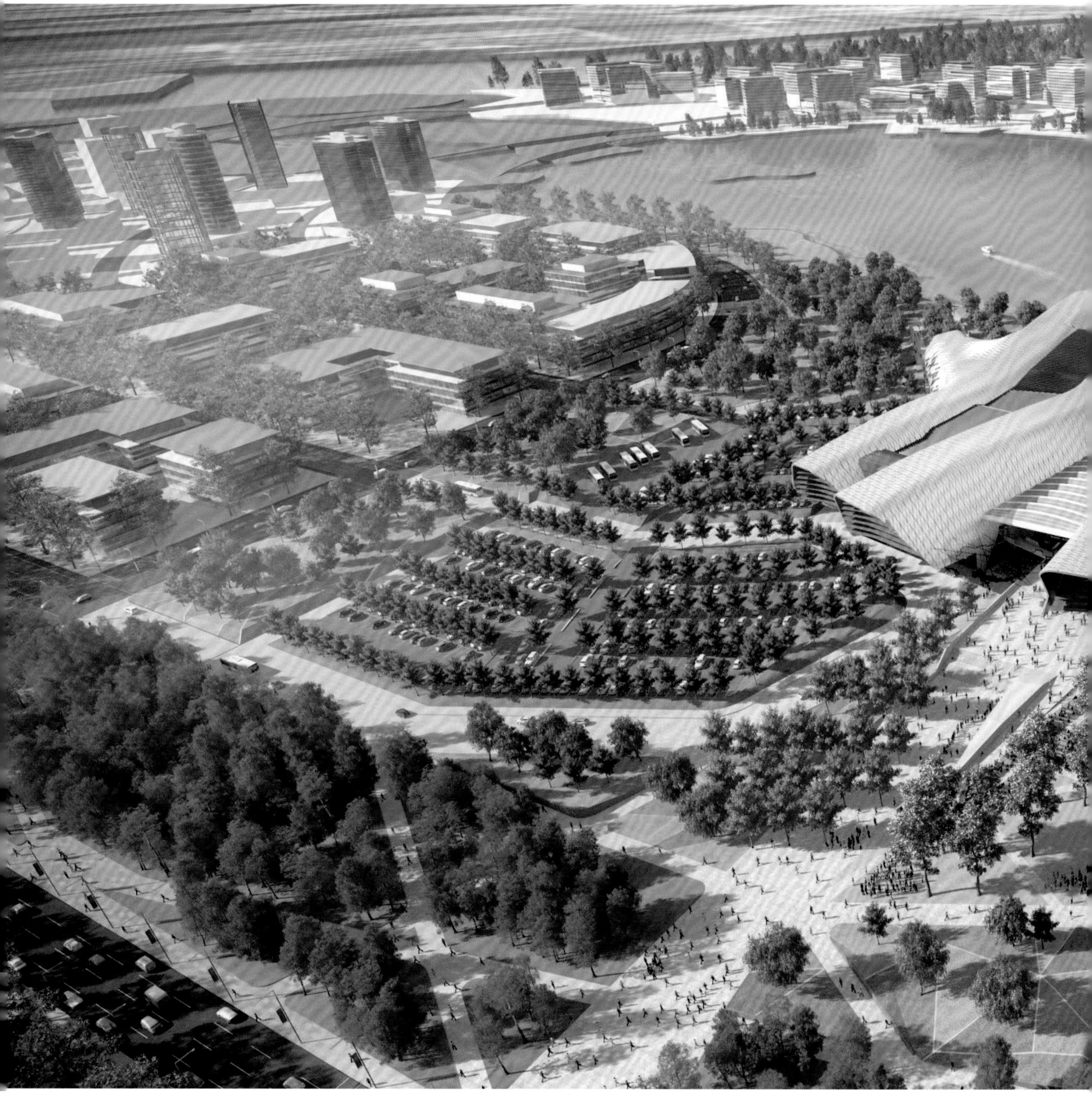

国家海洋博物馆设计方案效果图 2

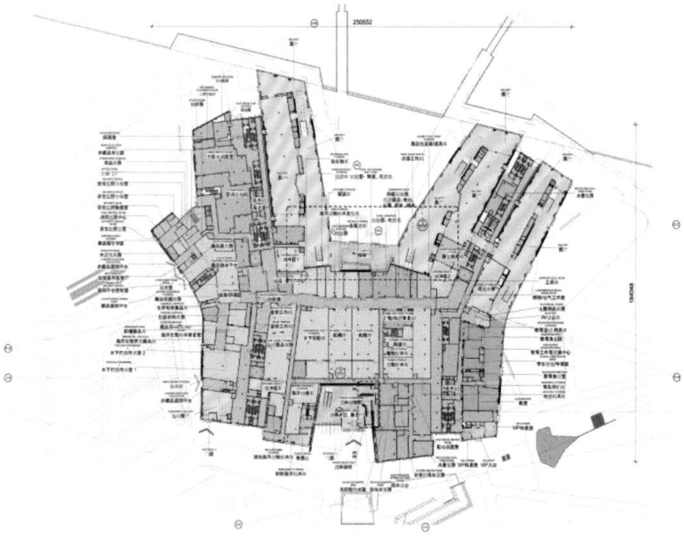

大沽标高 7 m 标高层平面图

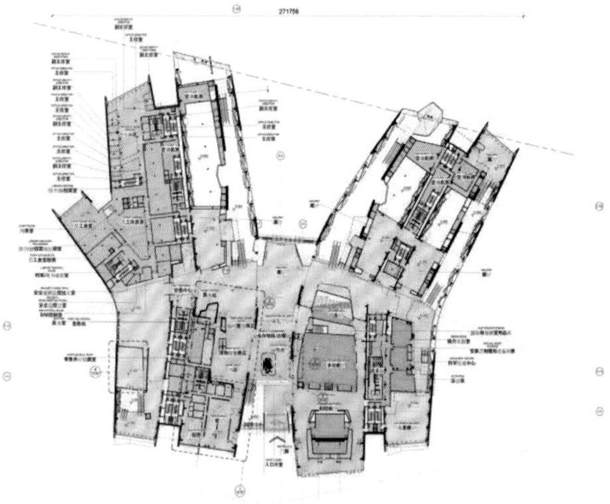

大沽标高 13 m 标高层平面图

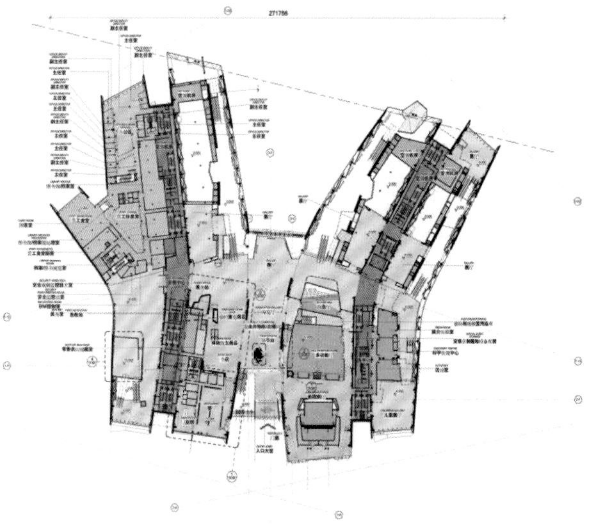

大沽标高 19 m 标高层平面图

大沽标高 7 m 标高层

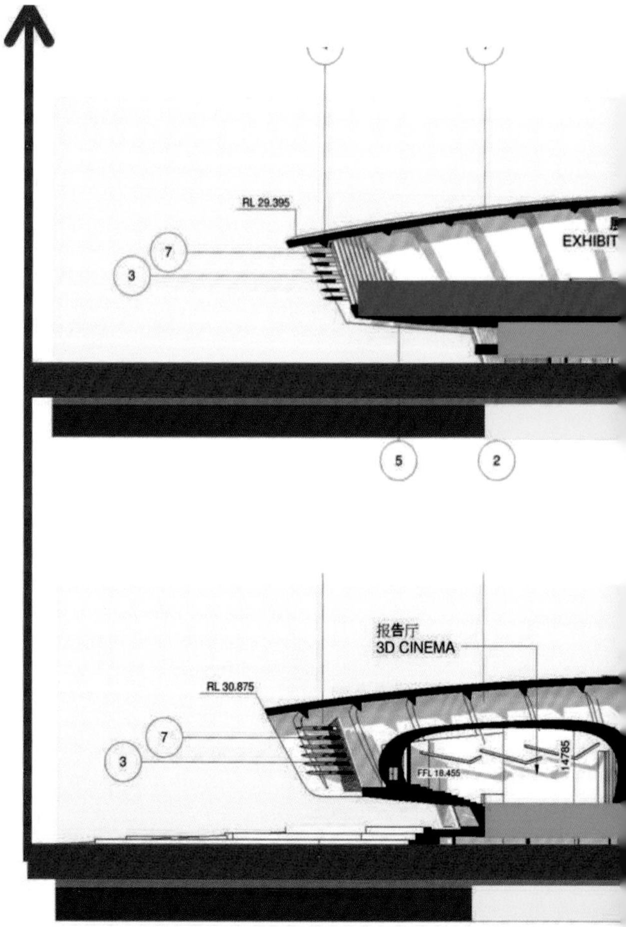

自 2014 年 10 月开始，国家海洋博物馆筹建办公室对项目的功能布局和展厅规模及数量进行了多次调整，并于 2014 年 11 月，对天津市建筑设计院调整后的设计进行了确认。根据国家海洋博物馆筹建办公室的要求，设计人员在该版图纸的基础上，进行了后续的施工图设计和各专项设计。

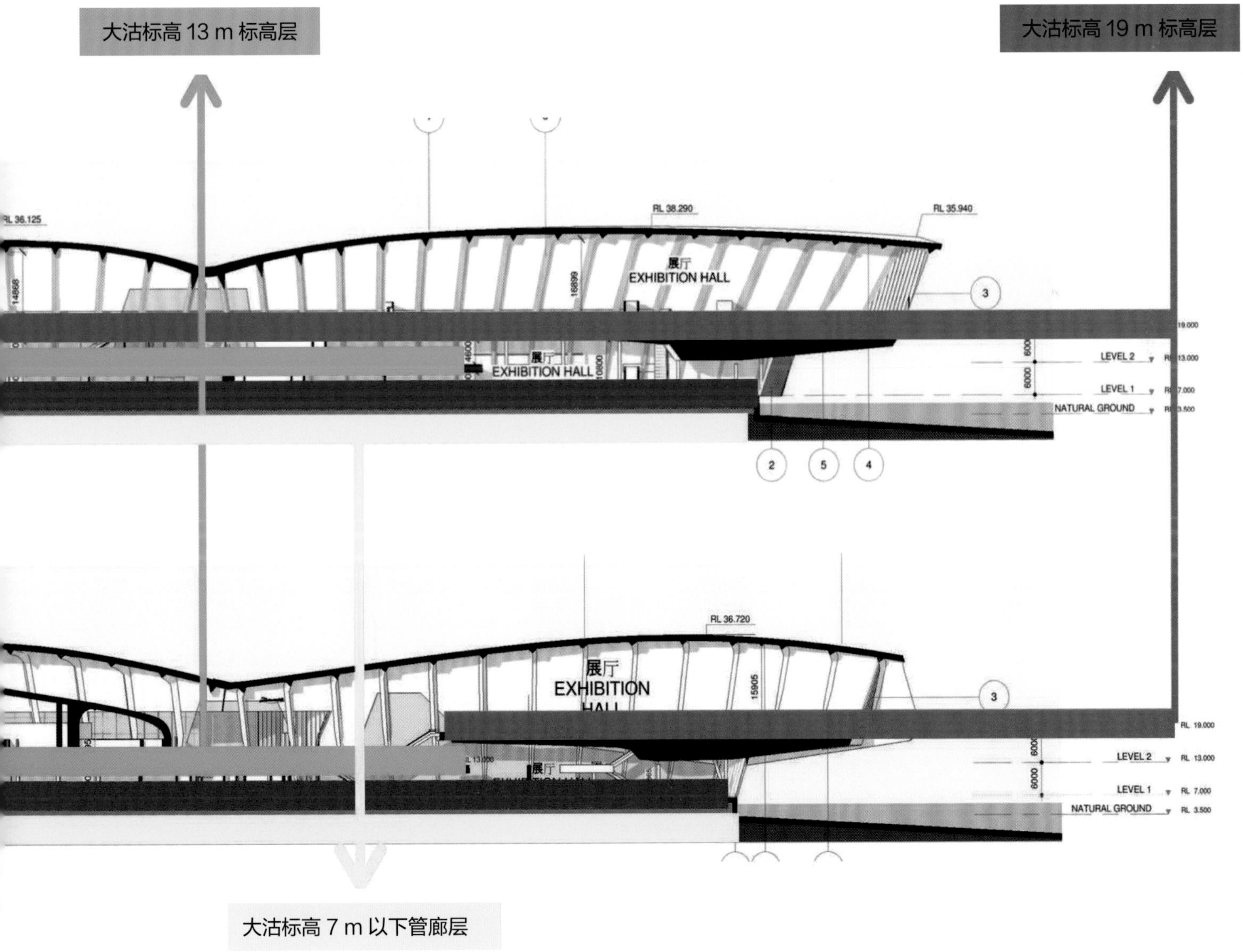

第二章　建筑设计

Chapter II Architectural Design

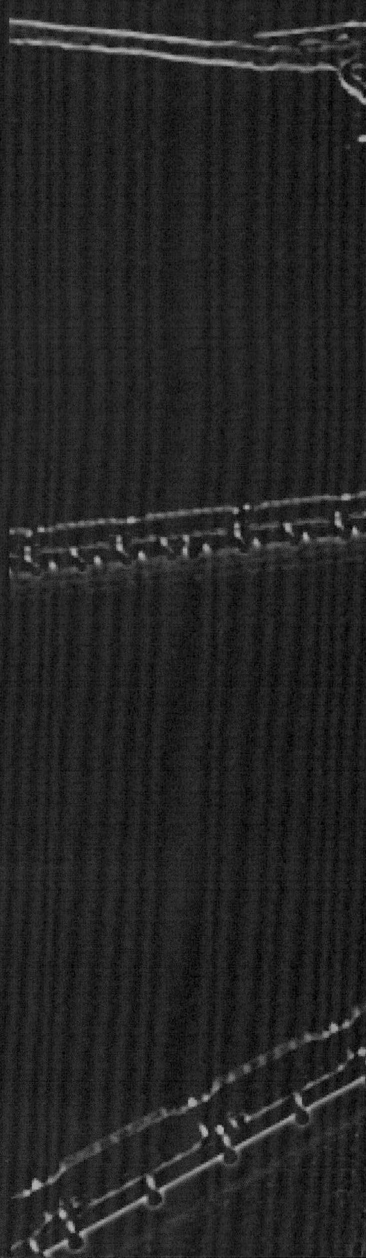

作为一座以海洋文化为主题的综合性博物馆，国家海洋博物馆的设计自始至终秉持着向海亲海、融合统一、挑战自我的设计理念以及国际视野与本土文化并重的设计原则。设计团队攻克了多项技术难题，最终呈现出一座能够充分体现中国现代建筑水准和建筑创作精神的国际化作品，其形神合度，将海洋与陆地文化交织的别样风姿展现得淋漓尽致，是科学与人文精神交融的成果，是对海洋建筑之美的生动诠释。

The design of the National Maritime Museum of China, a comprehensive museum on the theme of maritime culture, has always followed the concept of facing the sea and being close to the sea, integration and unification, and challenging itself, as well as the principle of balancing an international vision and the local culture. By overcoming a number of technical problems, the design team has finally presented an international masterpiece able to fully reflect the modern architectural standards of China and express the spirit of architectural creation. With integrated form and spirit that vividly shows the unique charm of the interweaving of maritime and land cultures, it is a result of the blending of scientific and cultural spirits and a vivid interpretation of the beauty of maritime architecture.

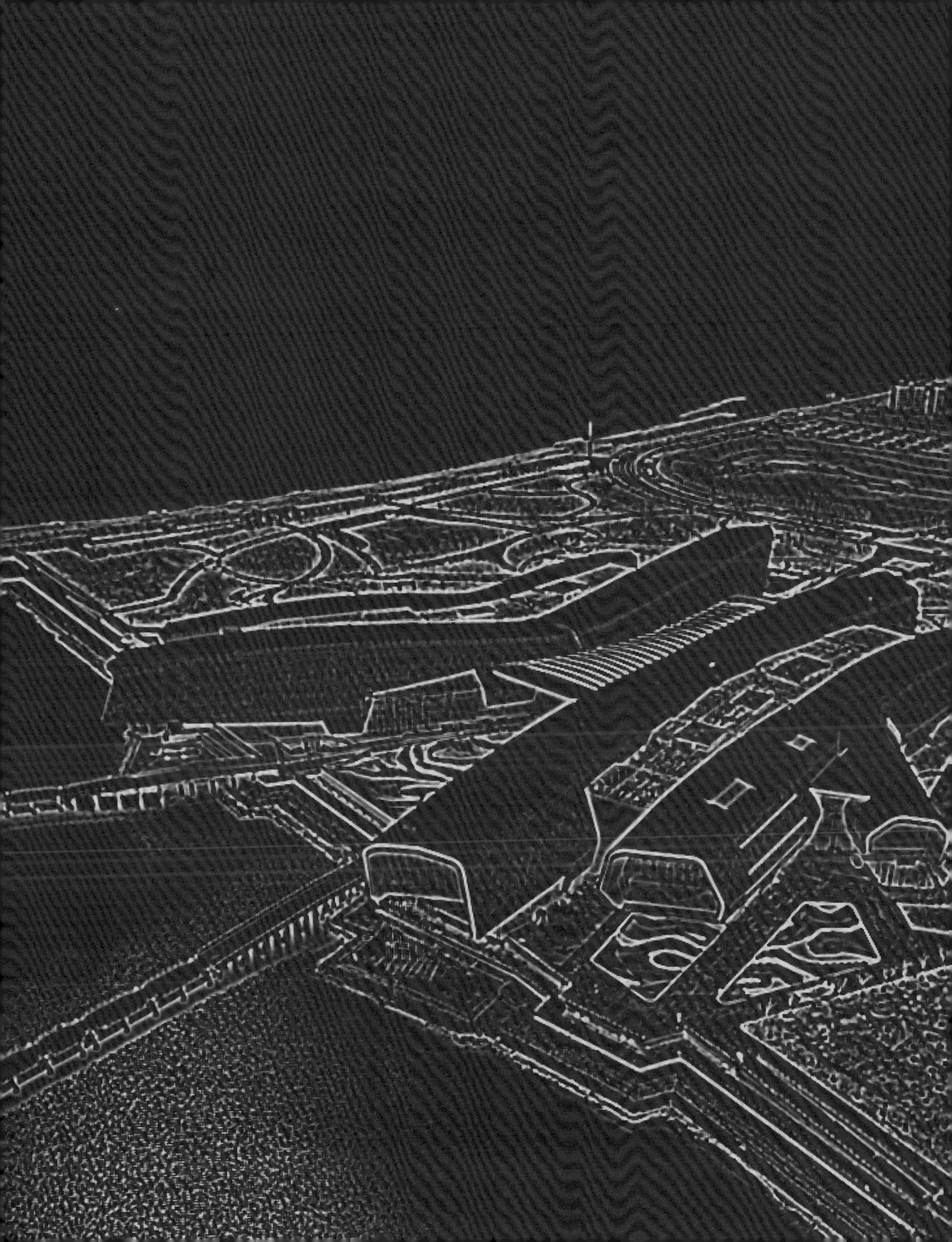

国家海洋博物馆是坐落在天津的第一座以海洋文化为主题的国家级综合性、公益性博物馆。该项目的建设旨在增强“中华民族海洋意识”，倡导“和谐海洋 · 和谐世界”的核心理念，体现了中国海洋发展战略，承担了重塑中国海洋价值观的重任。它不仅是天津市一座全新的文化地标，更是中国海洋事业发展的文化里程碑，建成后将构成整合滨海生态新城、南湾公园以及海洋自然环境的一个界面，并成为人类与海洋互动的新空间、新场域的媒介。

一、工程概况

1. 地理位置

项目位于天津市滨海新区中新天津生态城渤海南湾，规划面积为 1 km^2 的南湾公园内。

项目用地为填海造陆区域。用地西侧为旅游区、东方文化广场、欢乐水魔方、碧桂园滨海城、天成国际温泉酒店、贝壳堤湿地公园等项目，用地南侧为宝龙欧洲城、渤海海洋监测监视管理基地等项目。

国家海洋博物馆区位图

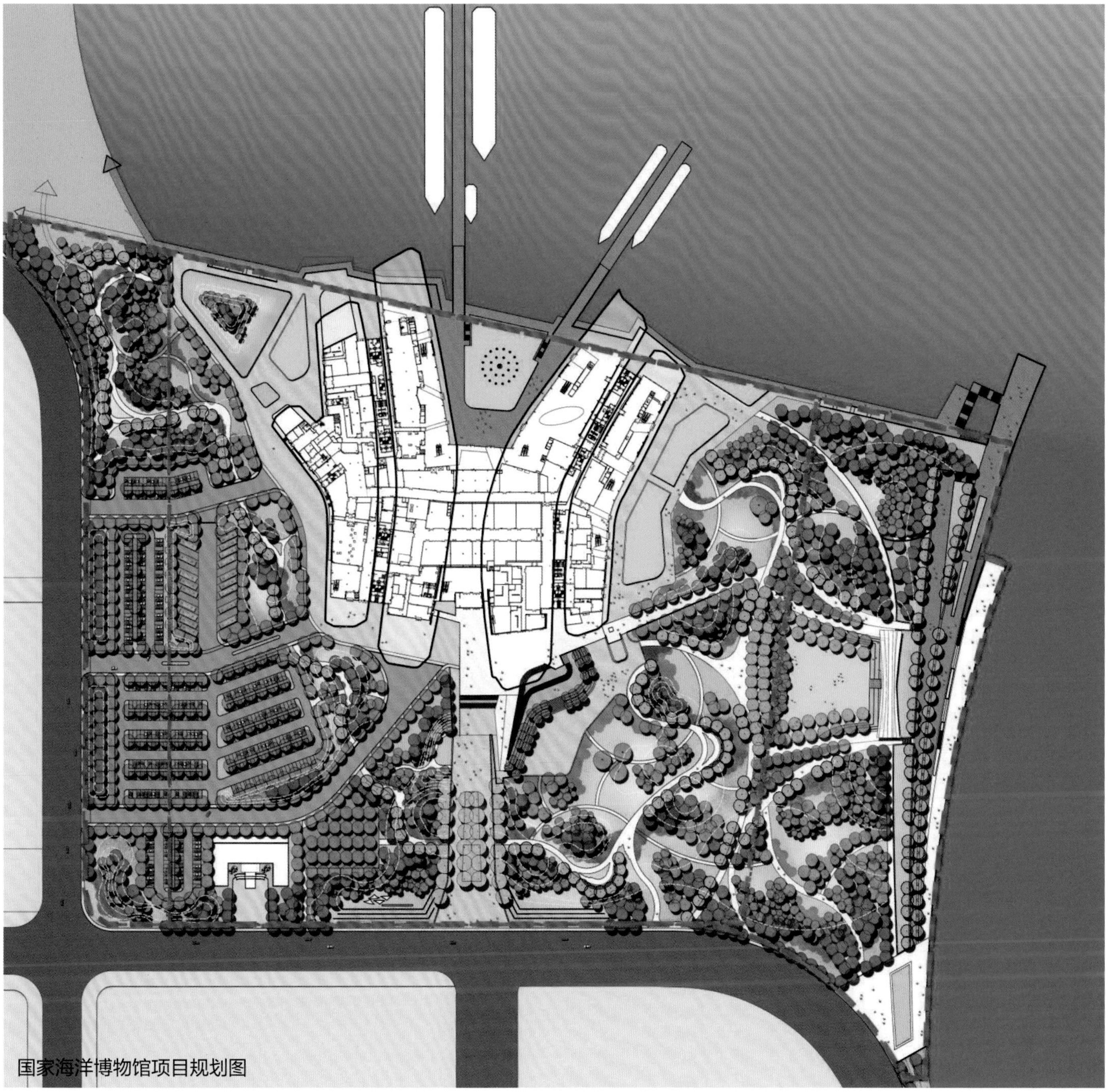

国家海洋博物馆项目规划图

2. 用地范围

用地面积：15 ha，后经东扩西进用地面积增至 23 ha。

总建筑面积：8 万 m^2，建筑处于大面积的公园绿地中，用地东侧和北侧临海，西侧和南侧临城市干道，交通出行顺畅便捷。

二、交通流线分析

1. 人车分流的交通架构

西面的荣盛路上分别设有社会交通车辆入口、员工车辆入口、藏品车辆入口三个机动车出入口。人员步行由南面的海轩道进入，贵宾车辆由东南向专用入口进入。

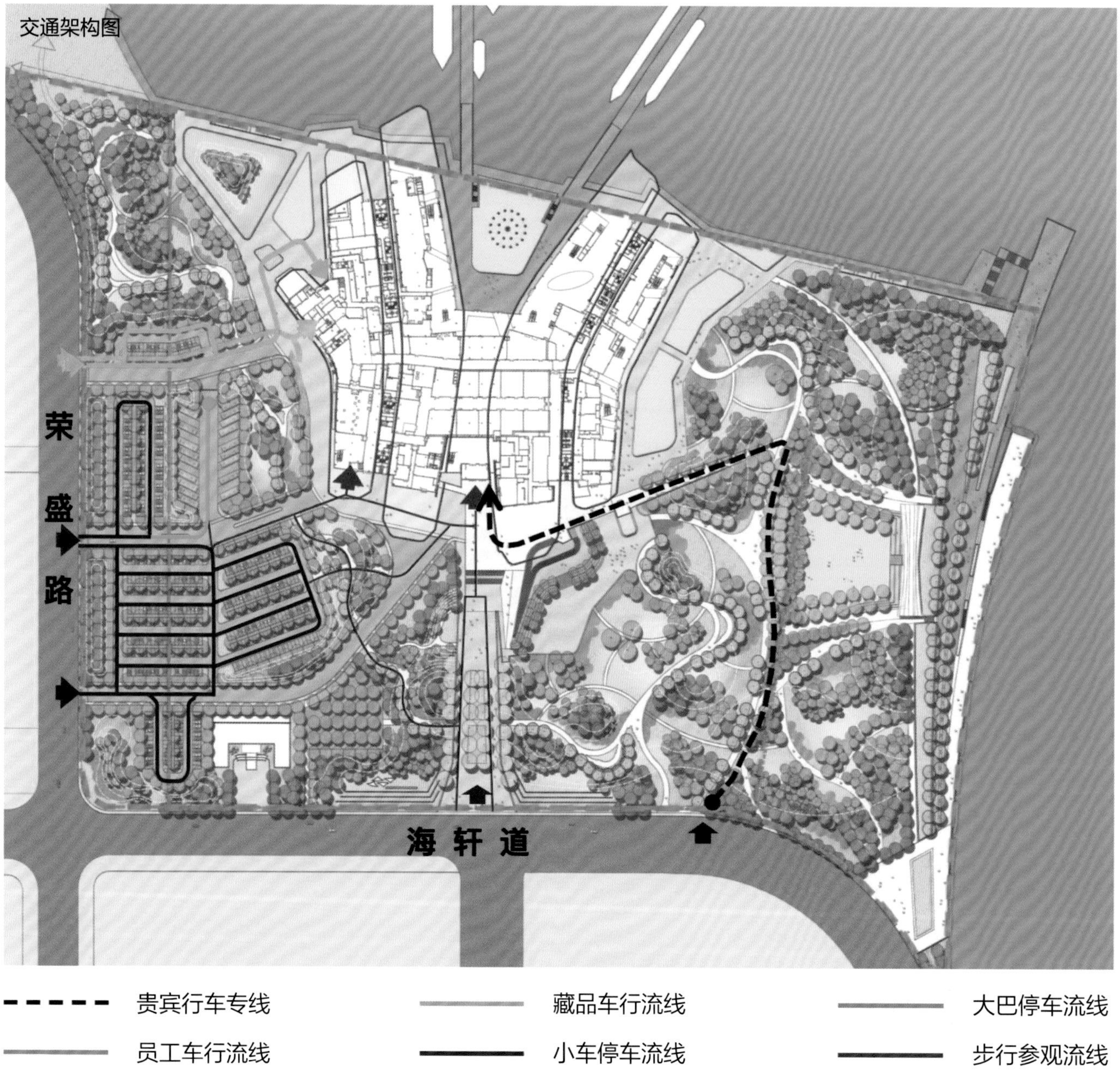

2. 步行游览的交通流线组织

在整个园区内，结合景观设计，设置四条步行游览交通流线，分别为：公园游览主流线、游览次流线、环馆游览流线和沿海游览流线。

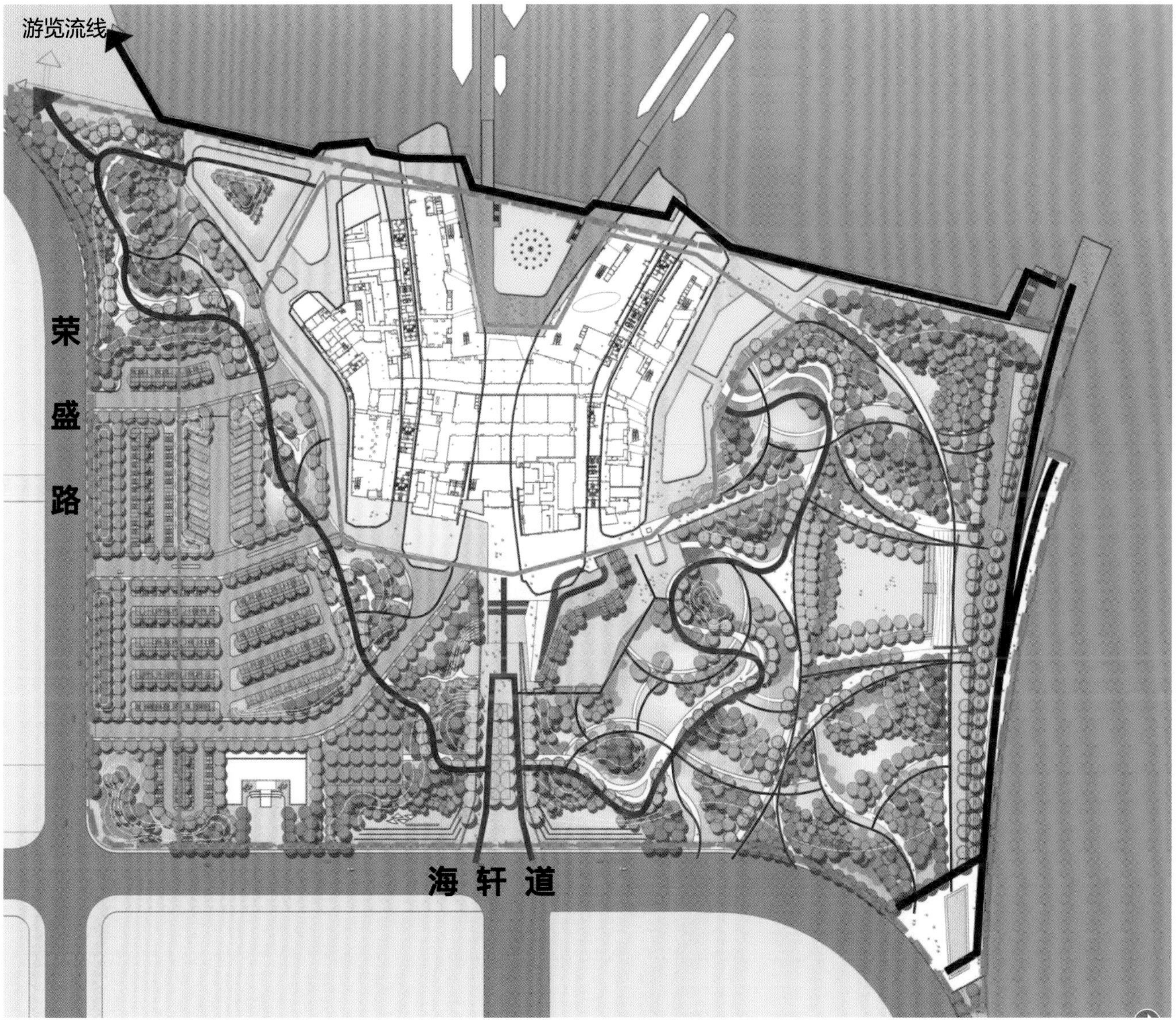

公园游览主流线　　环馆游览流线

公园游览次流线　　沿海游览流线

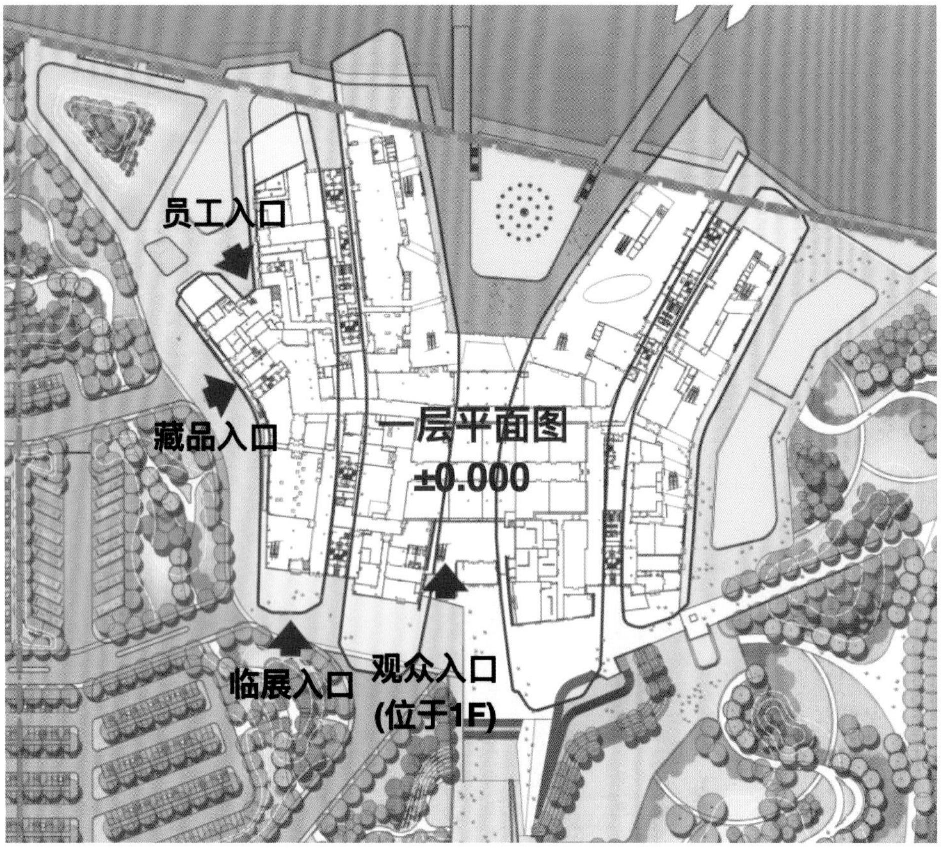

3. 人流复合式入馆流线组织

鉴于安检制度的程序要求，将一般观众和贵宾进馆入口分设于一层广场和二层平台。一般观众从一层广场进入安检大厅，乘自动扶梯到二层中央大厅，一层广场上方有700 m^2的二层平台，可为排队等候入馆的观众提供遮阳避雨的等候空间。贵宾入口设在二层平台上，贵宾通过专用交通通道直接进入中央大厅。

展馆入口位置示意（组图）

贵宾入口位于博物馆主入口的二层平台之上。员工入口位于西北向和西向。

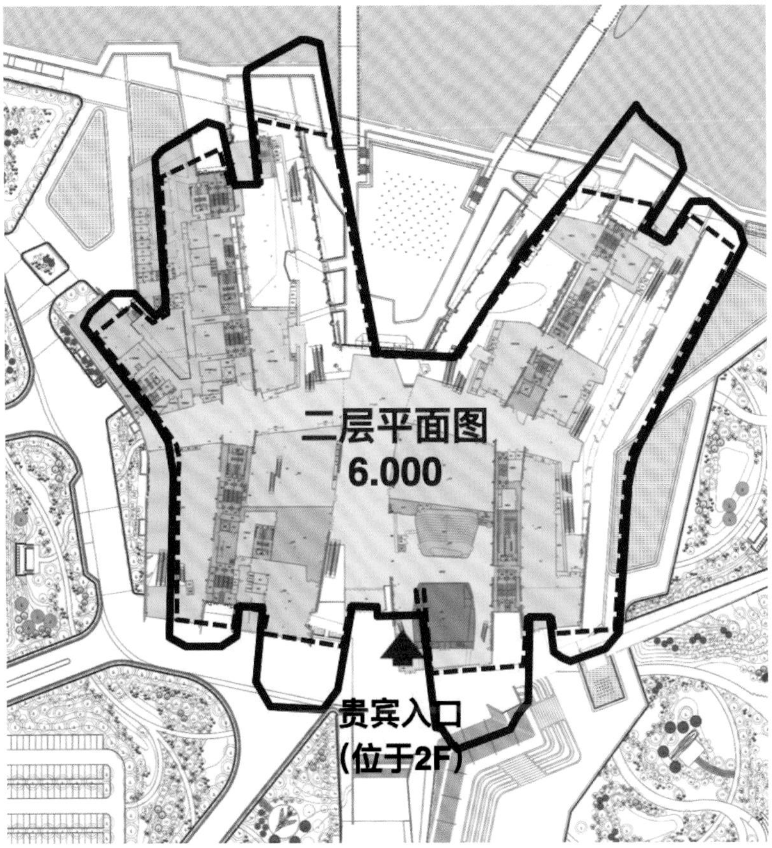

三、“馆园融合”规划架构

“馆园融合”是总体规划设计布局的大思路，设计将建筑与景观和场所整体营造，实现了从馆舍天地走向大千世界的广义博物馆的理念。

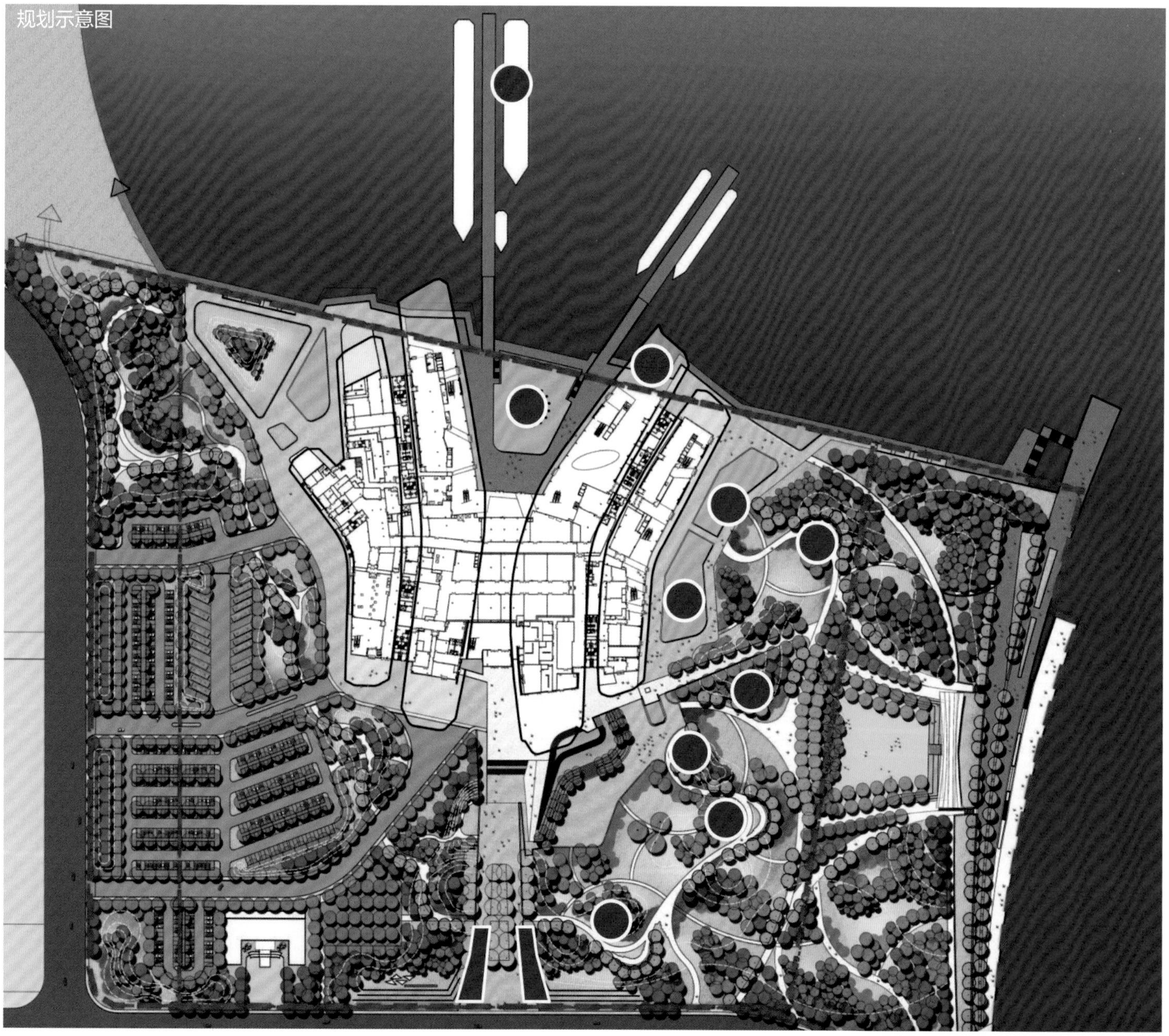

规划示意图

1. 架构陆路与水路空间交织

海上俯瞰

陆上俯瞰

海陆环境总览 1

海陆环境总览 2

园区陆路与海路的空间交织

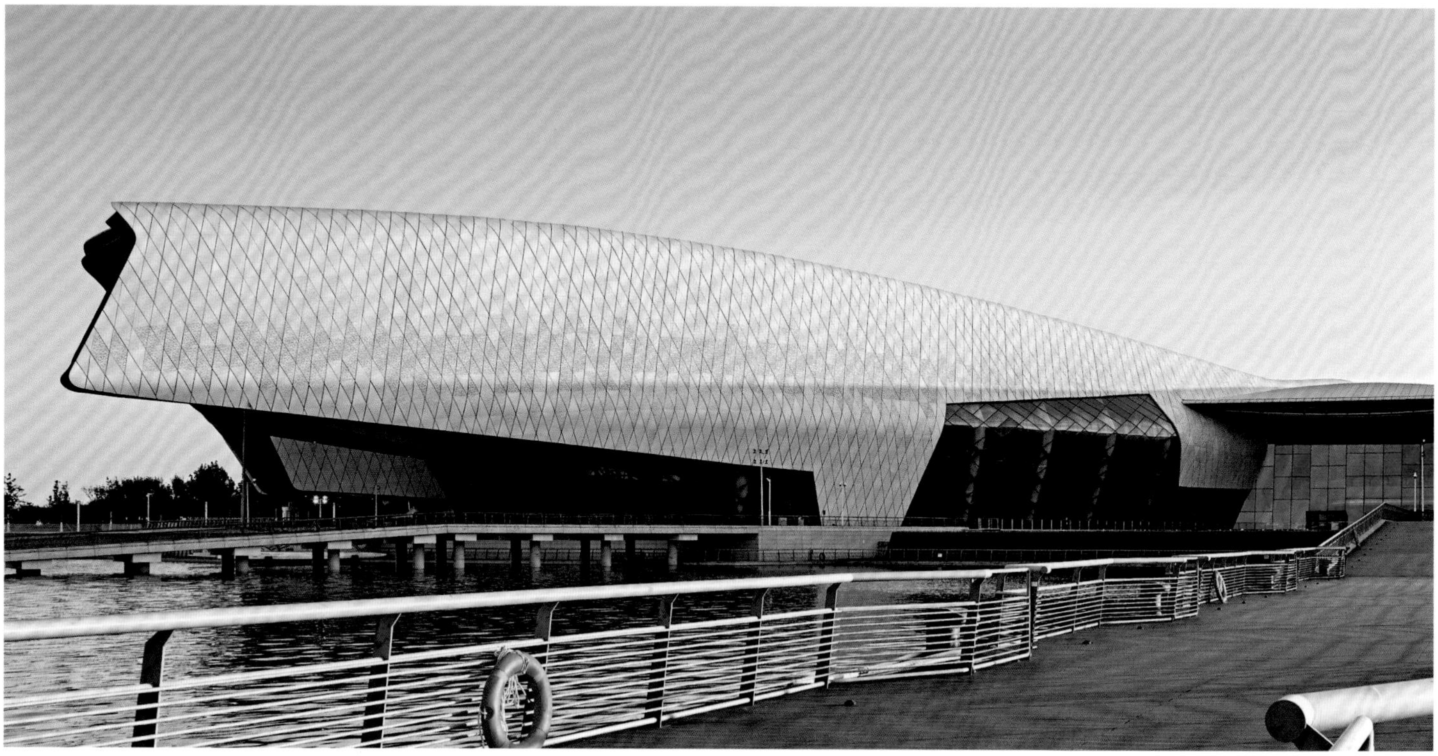

建筑主体向海侧延伸（组图）

建筑周边水体布置（组图）

2. 建筑与景观空间的无缝对接与融合

（1）入口空间

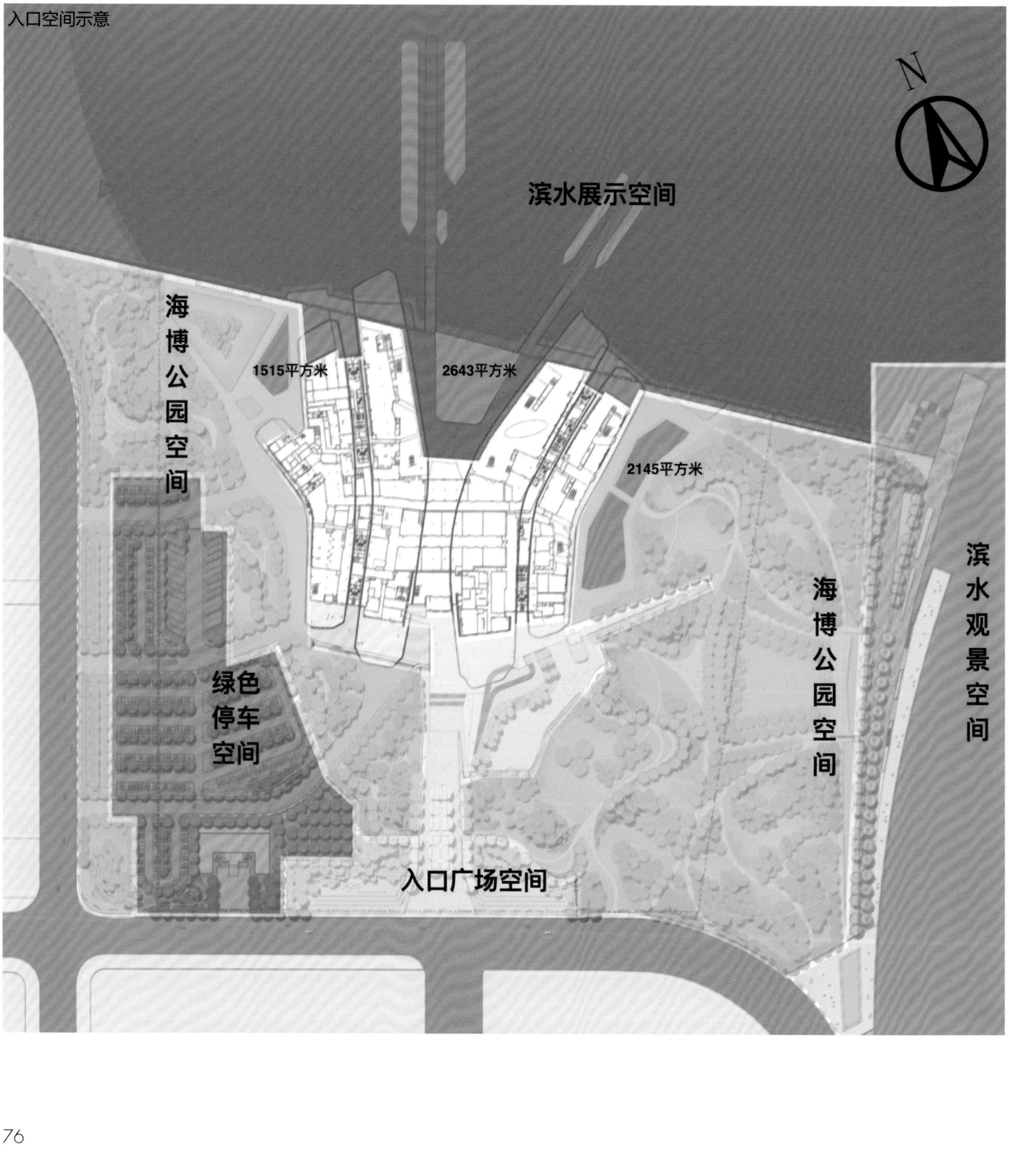

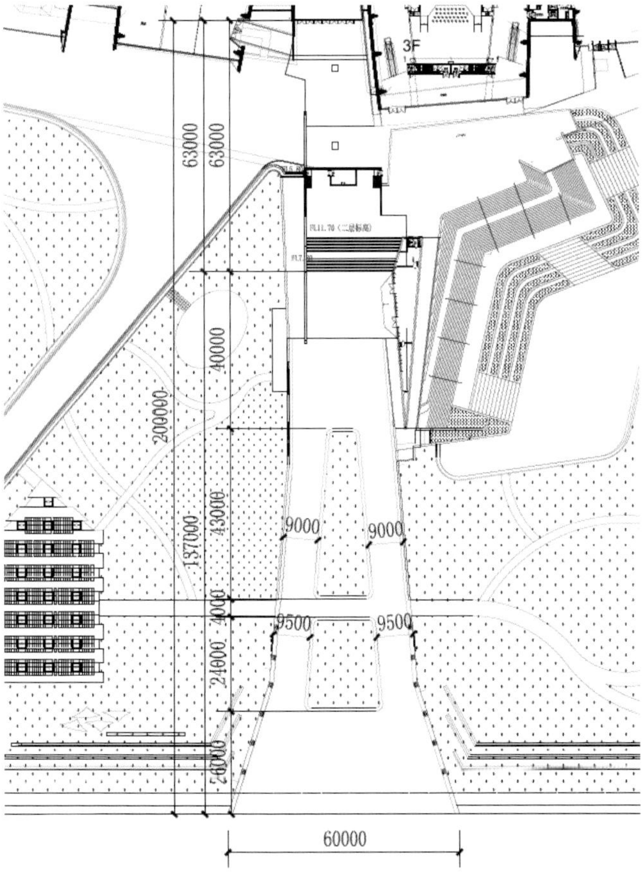

入口广场平面

园区入口（组图）

建筑北侧

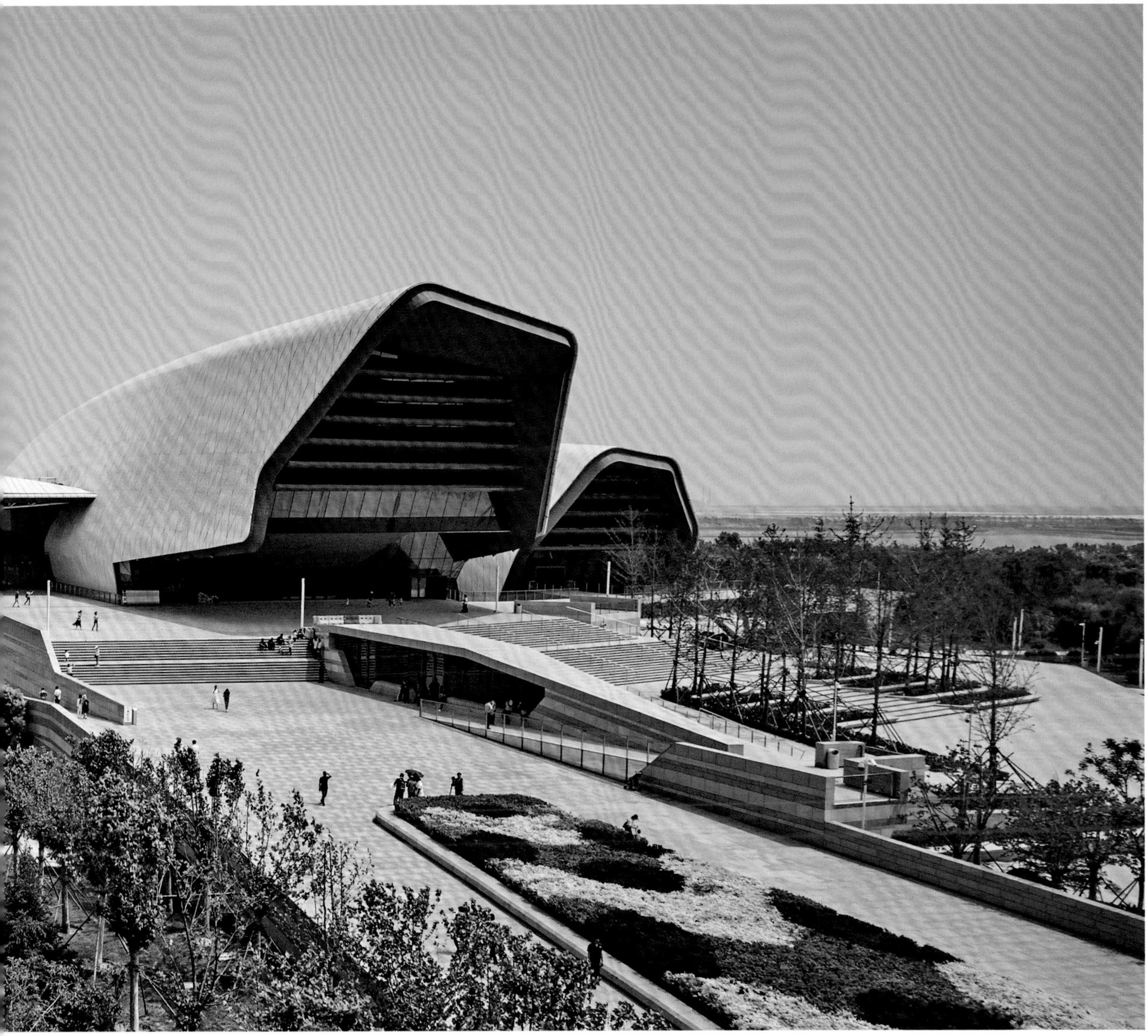

入口侧挑檐 1

（2）建筑入口空间

建筑入口空间

入口侧挑檐 2

入口侧挑檐 3

（3）海博公园空间

园区展示（组图）

（4）绿色停车空间

绿色停车空间

便捷的绿色停车场

（5）滨海展示空间

滨海景观（组图）

（6）滨水景观空间

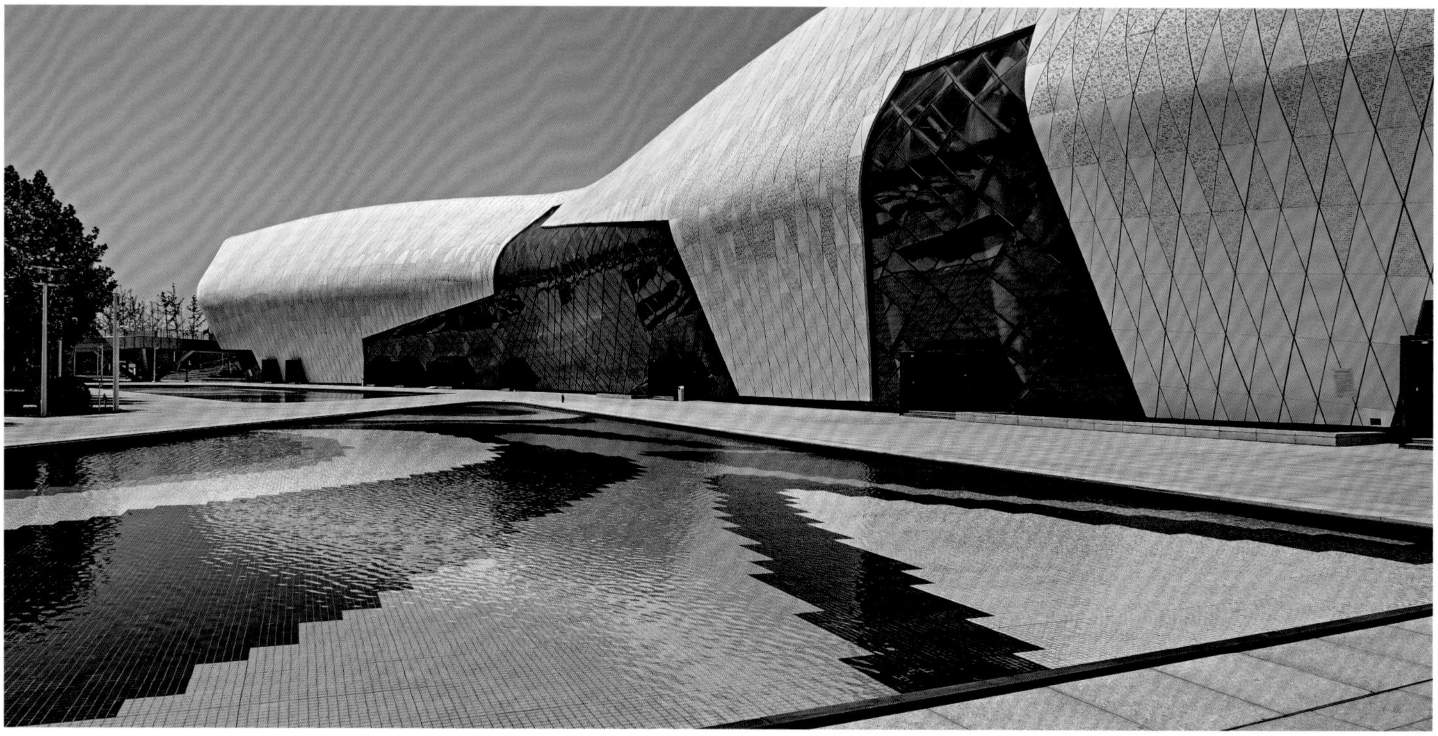

滨水景观（组图）

3. 外展与内展场所一体化

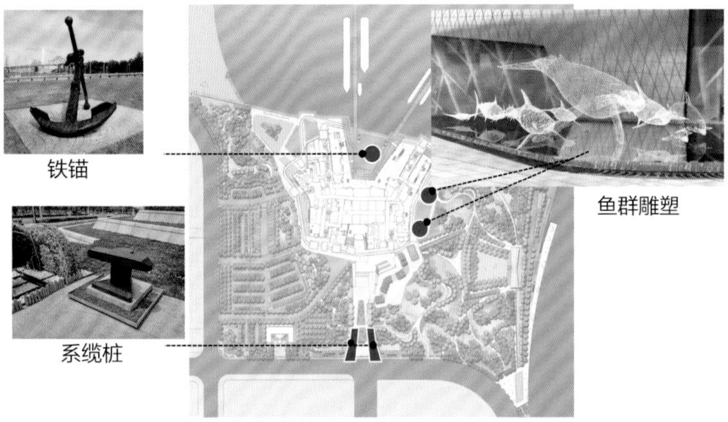

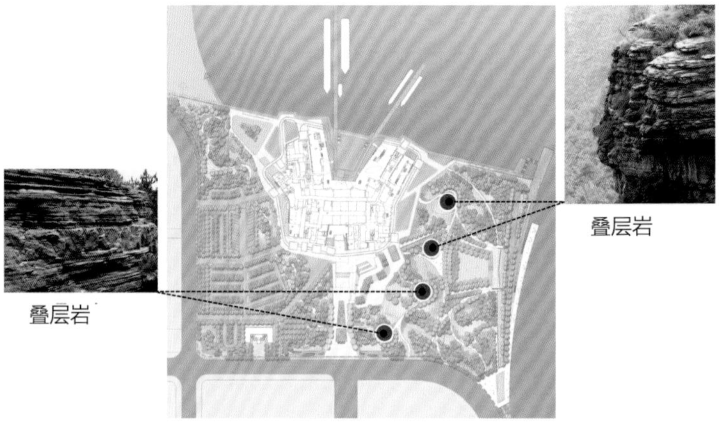

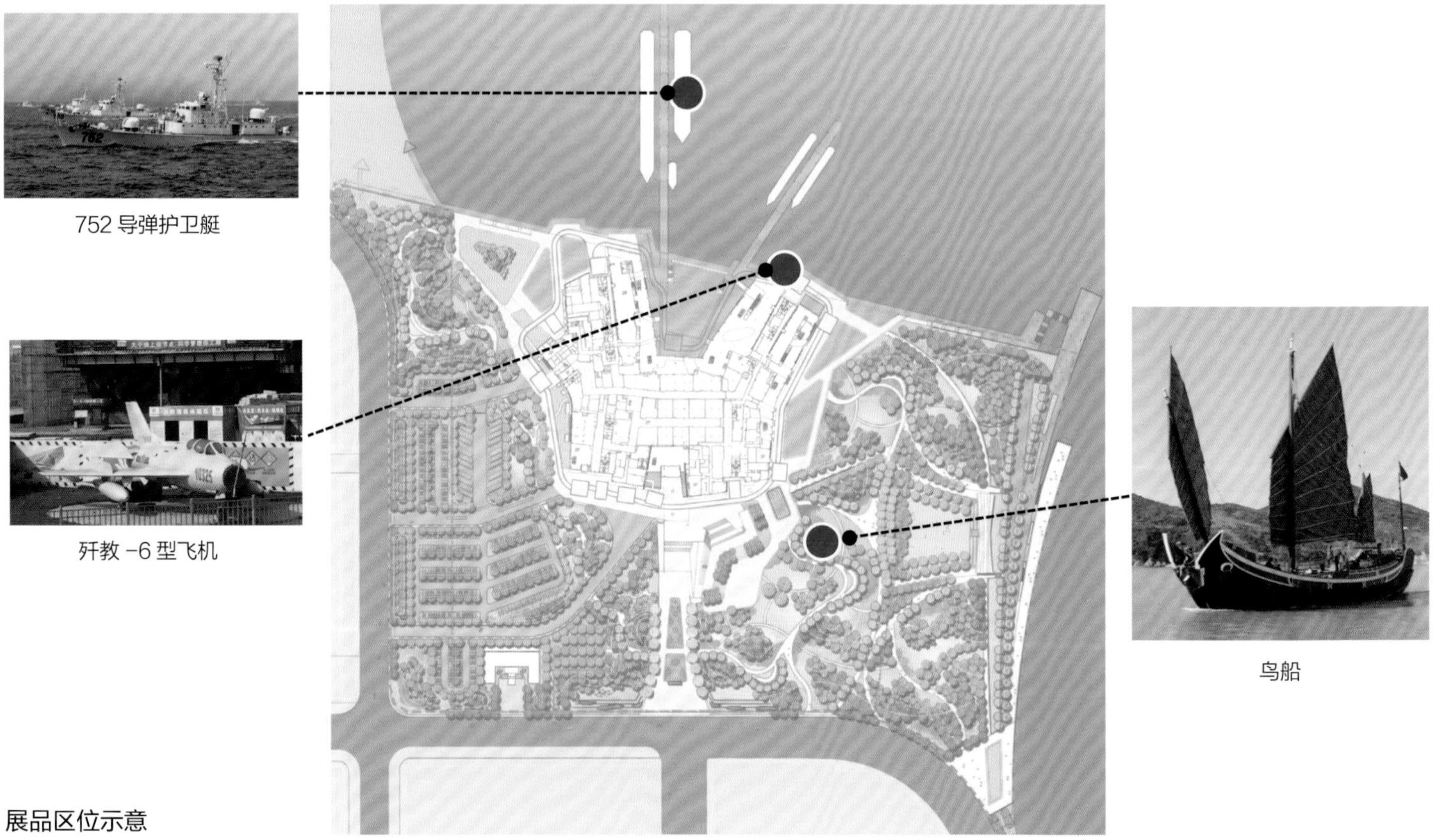

展品区位示意

752 导弹护卫艇（组图）

752 导弹护卫艇

大溪地号木质帆船（组图）

歼教 –6 型超音速喷气式歼击教练机

四、开放发散的设计理念

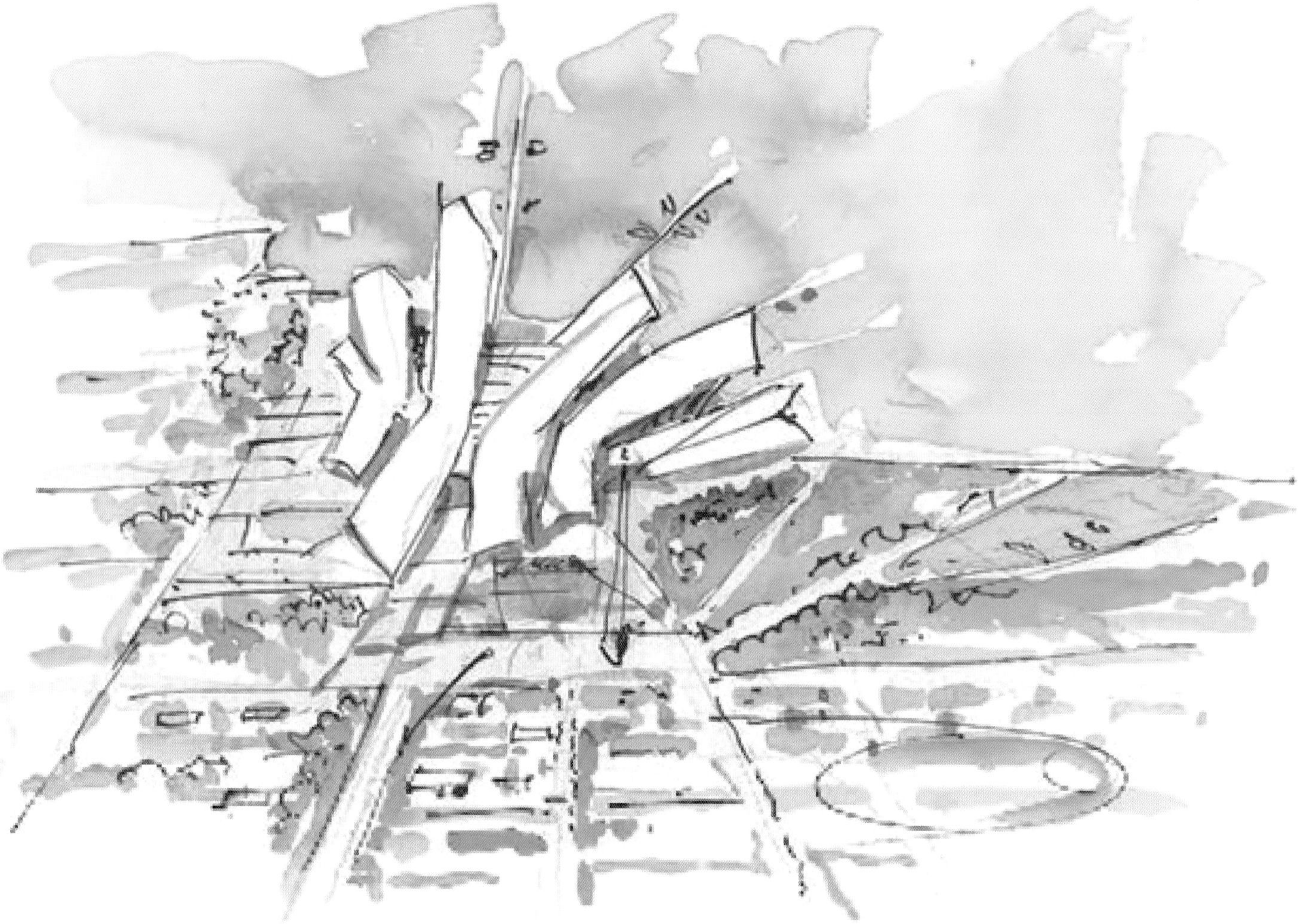

“开放形”的建筑形态是“近海临都”博物馆的个性，是设计创意的核心理念，设计运用隐喻的非线性手法，呈发散形由陆地向海洋展开，架空在海面上，以表达以下特质。

●向海亲海——建筑品格的特质，展示建筑架构与海洋的空间交织，展示陆地文明向海洋文明的探索和追求。

●融合统一——以中央大厅为基准空间由内向外发散状的建筑形态，展示建筑空间与多元展陈主题艺术空间的融合，展示展览空间交通组织流线的外在体现，展示建筑形式与功能的完美融合，彰显空间结构力学与建筑美学的高度统一。

●挑战自我——陆地文明向海洋文明探索追求的进程，展示“建筑人”不断挑战自我，攻克多项技术难题，以见证陆海文明载体落成的全过程。

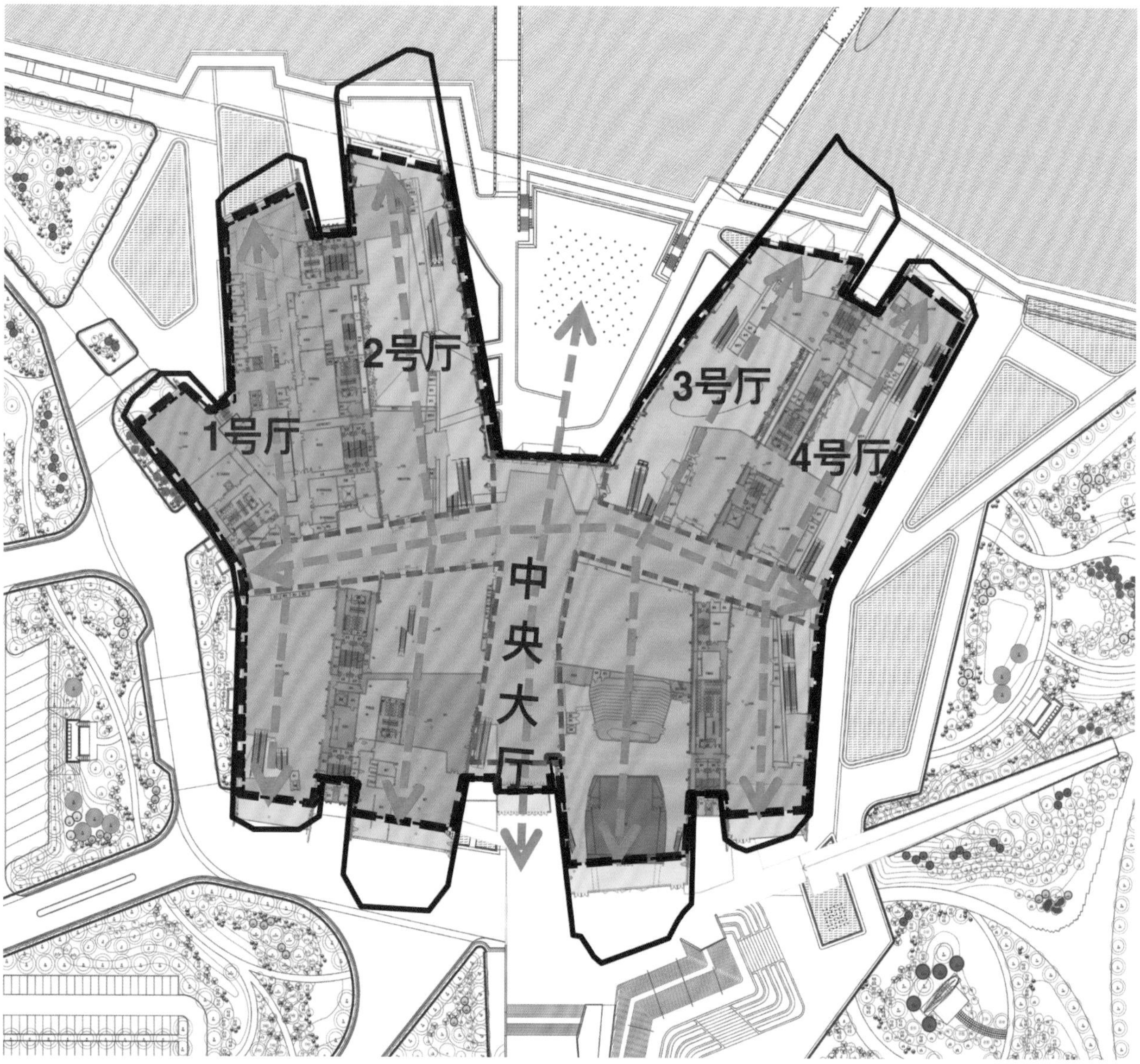

“开放形”的建筑形态

对“开放形”的建筑形态，仁者见仁，智者见智，有人说它像“跃向大海的鱼群”或“停泊岸边的船只”，又像“伸出张开的手掌” 以至“海洋的多种生物”等等。总之它柔美而不具象，并具有启发性。通过流畅动态的建筑语汇，令参观者对博大精深的海洋文化元素产生无限的遐想，完成对建筑的认知。

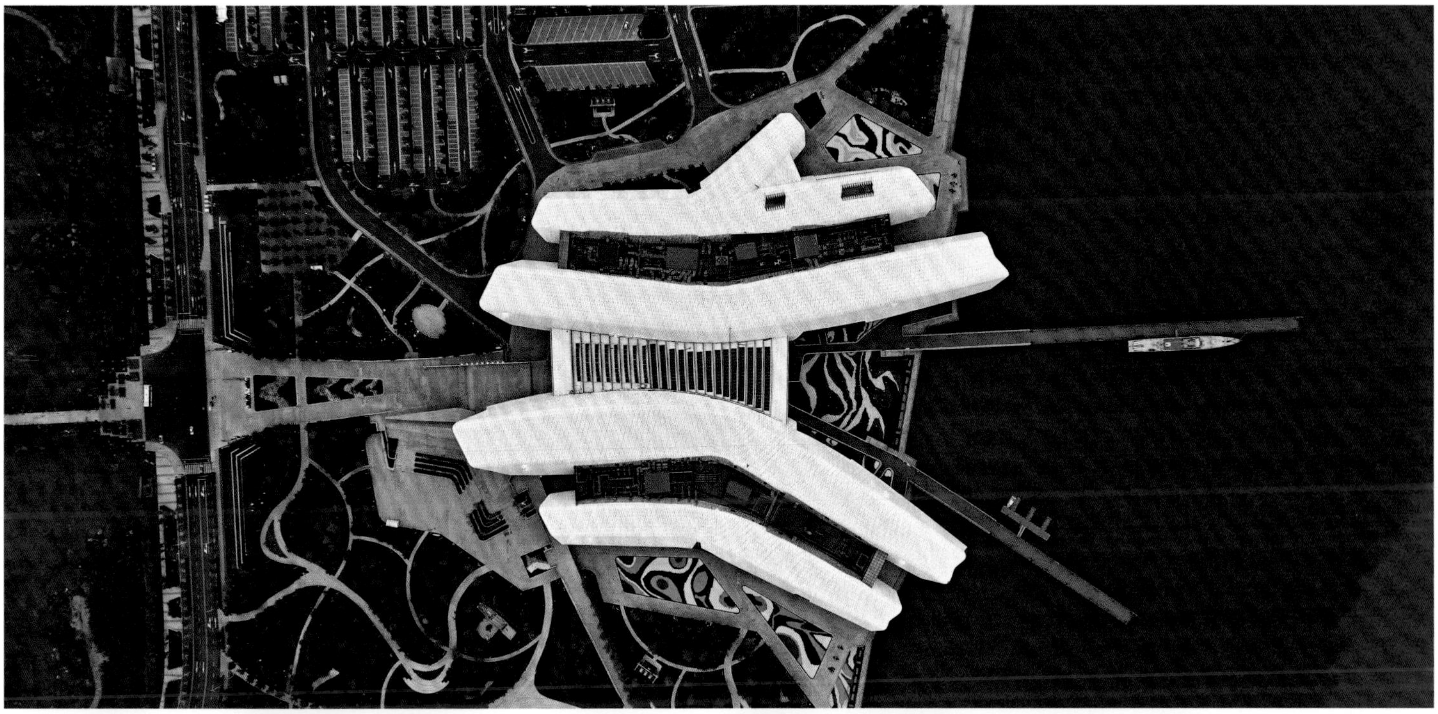

张开的手掌　海星　鱼群　海葵　港中的船只　珊瑚 / 海螺

建筑意向的表达

技术图纸－功能分区

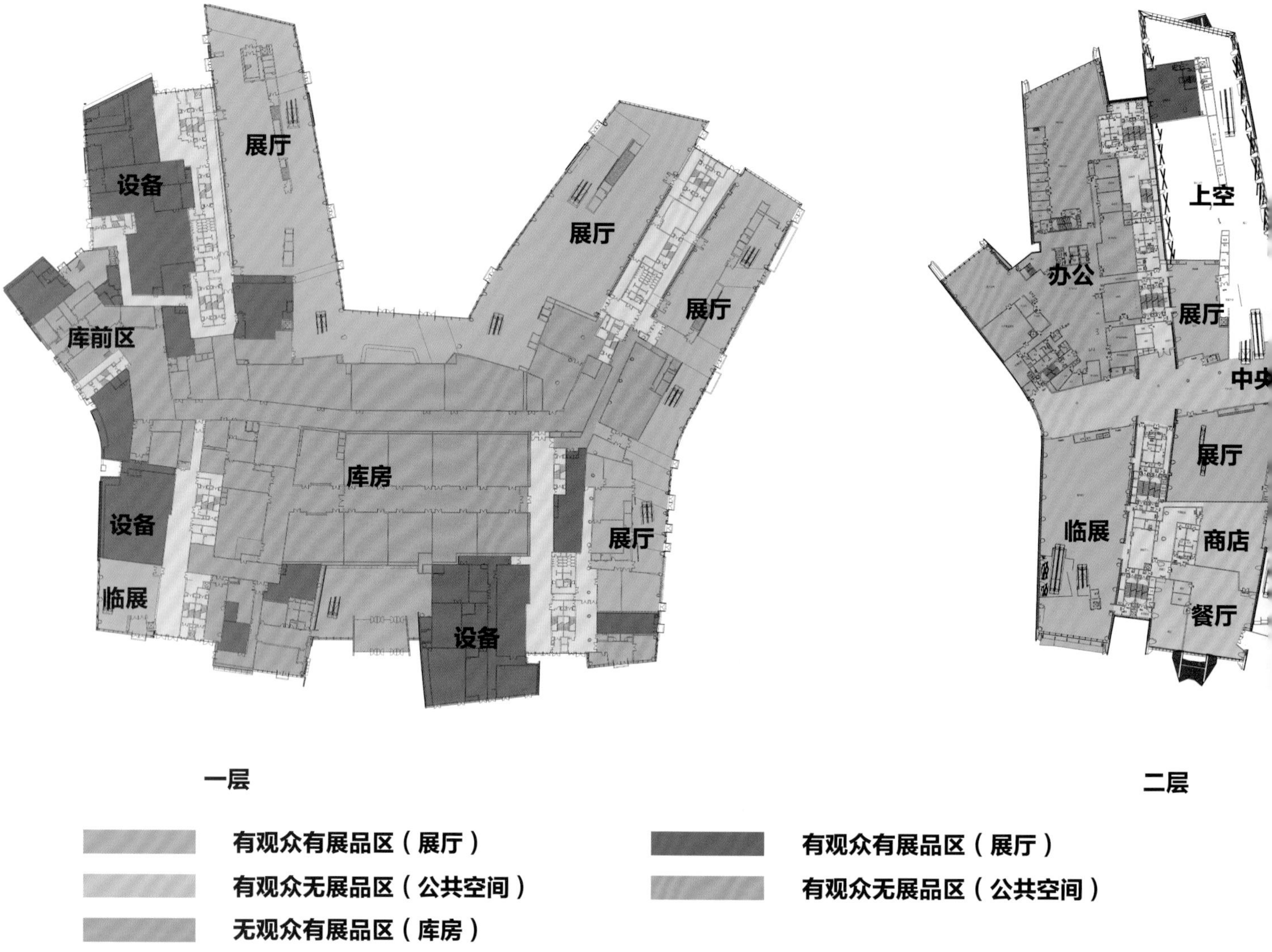

各层功能空间示意（组图）

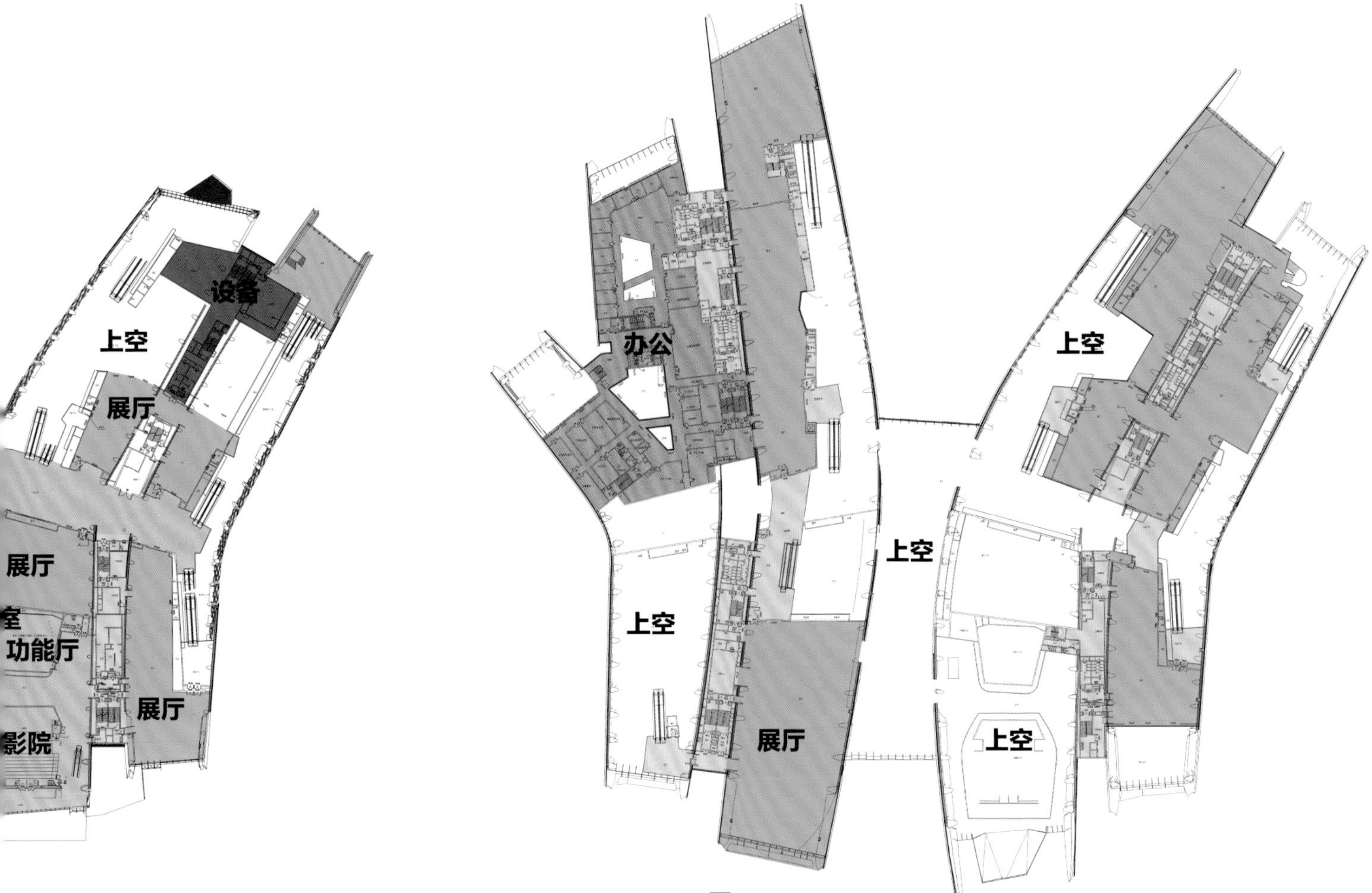

三层

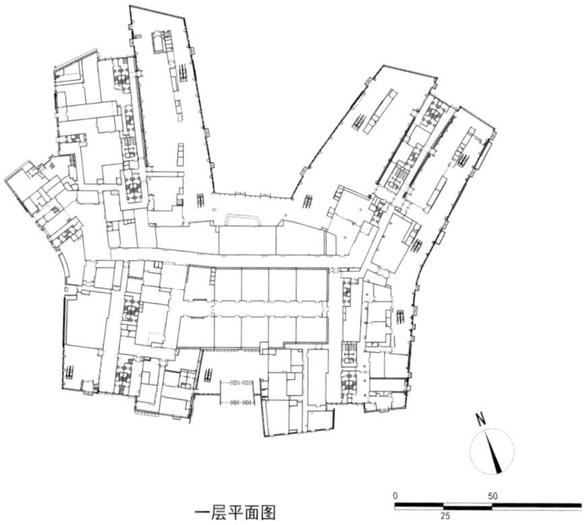

一层平面图

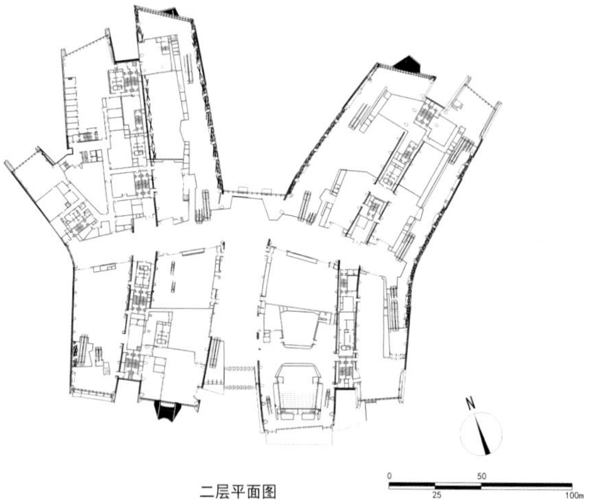

二层平面图

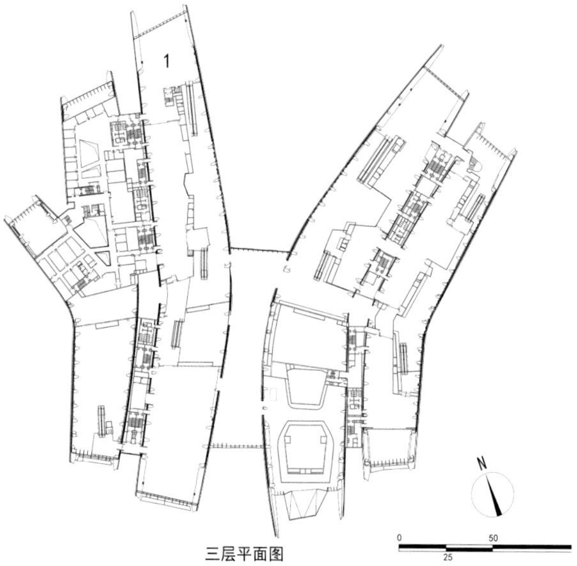

三层平面图

各层平面图（组图）

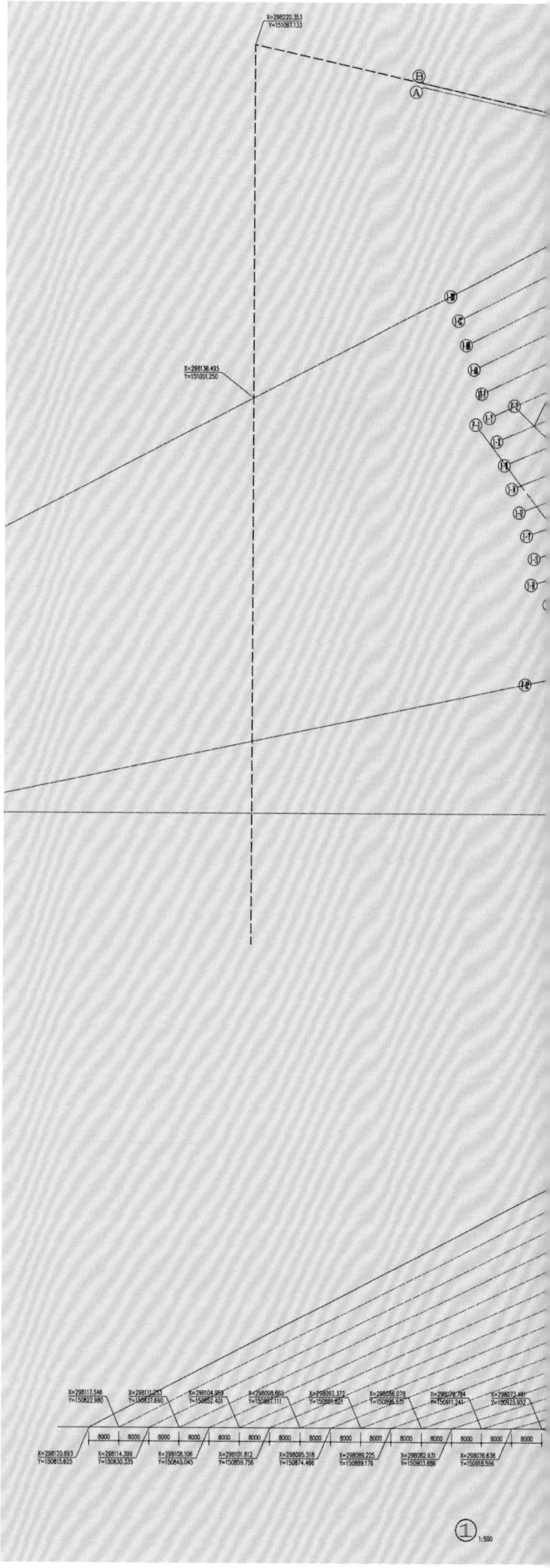

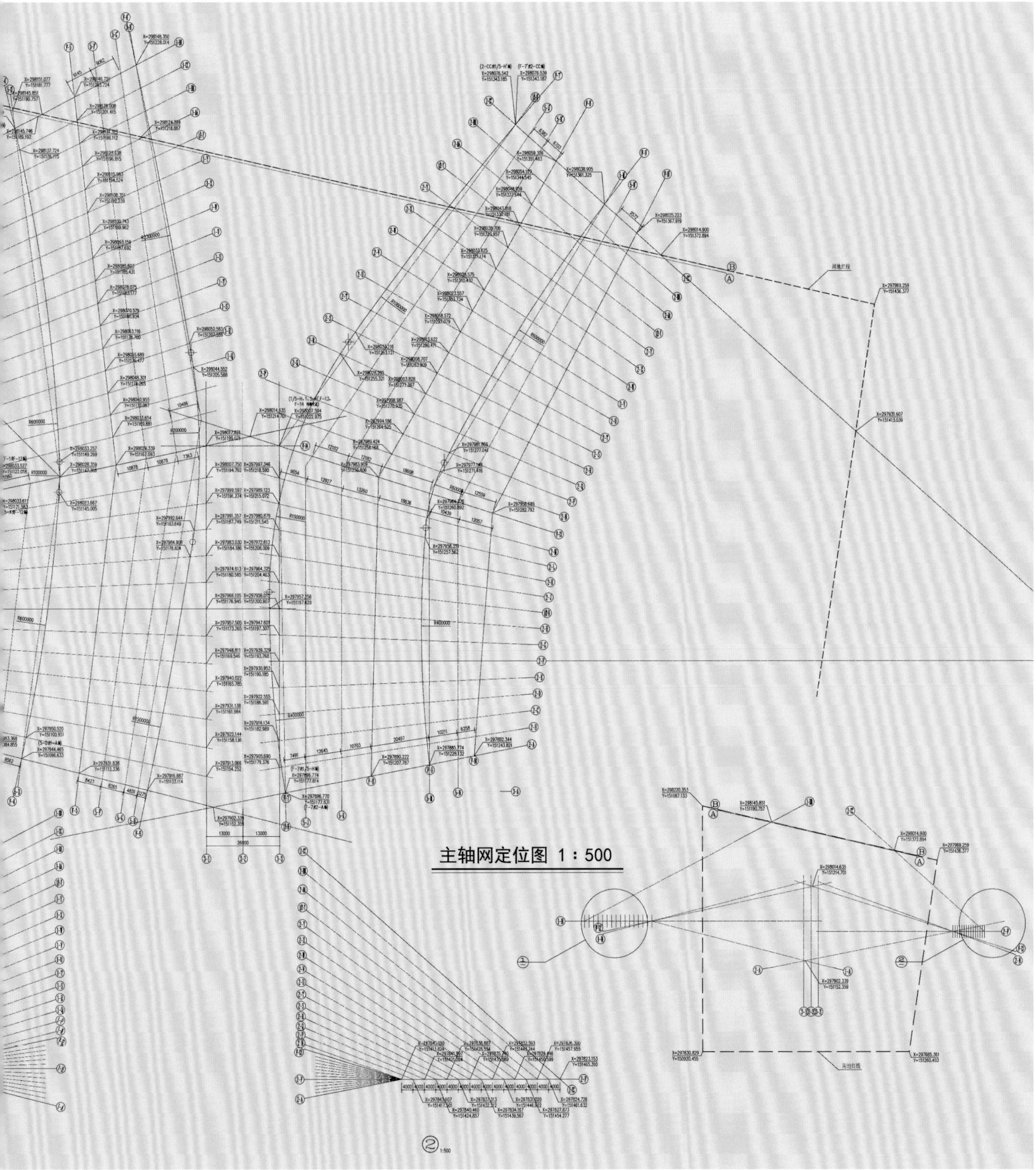
主轴网定位图 1：500

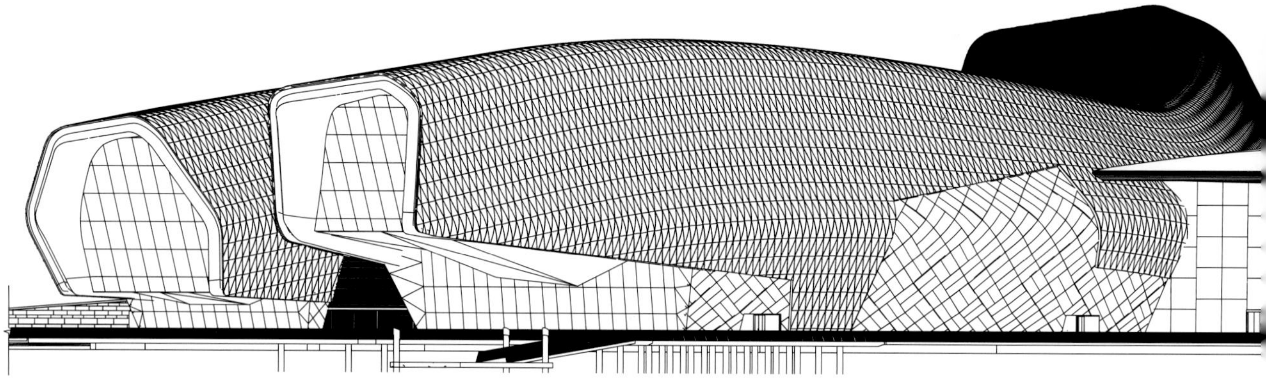

北立面

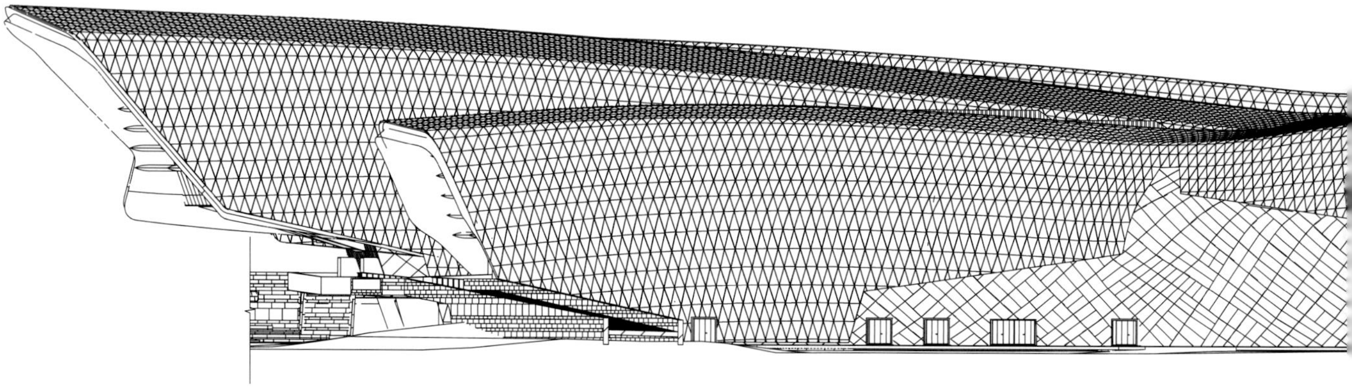

东立面

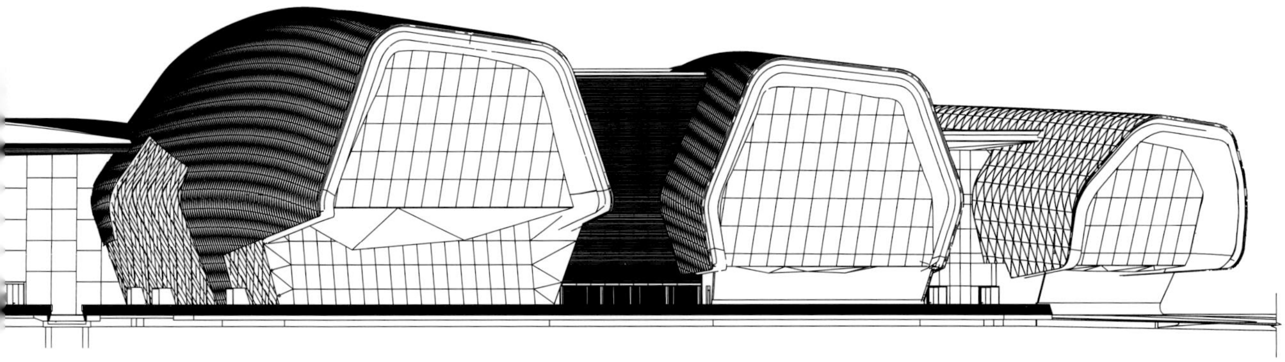

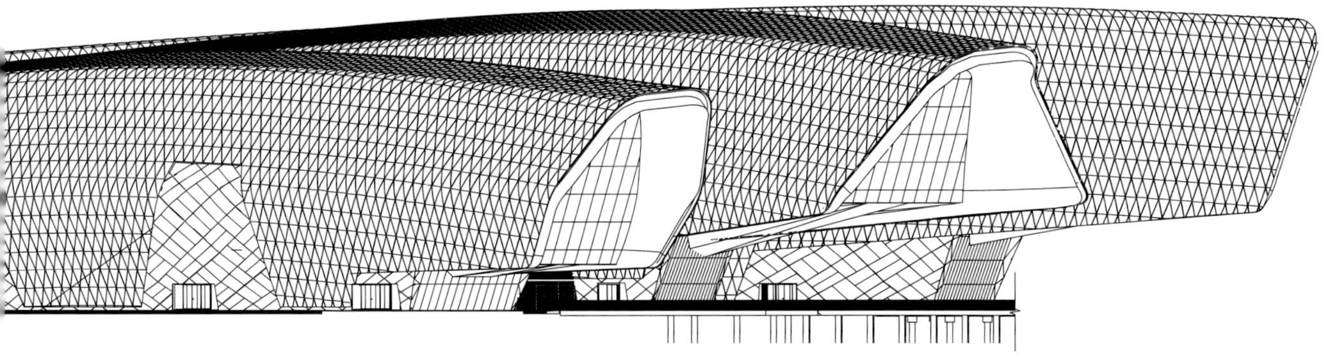

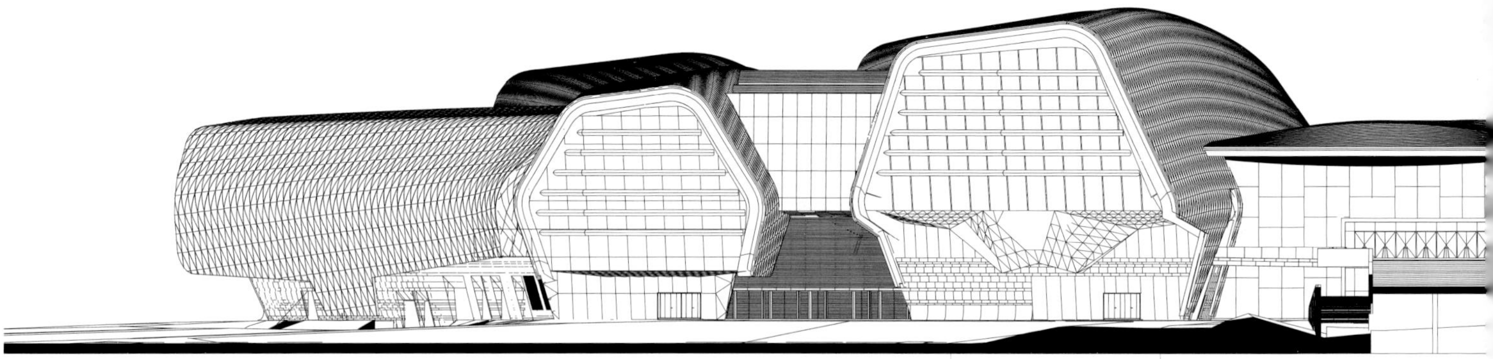

南立面

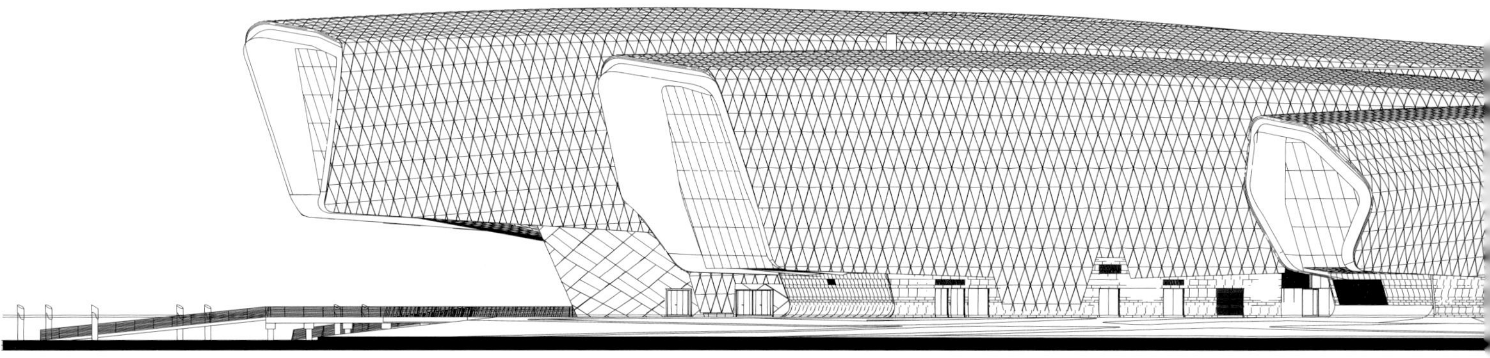

西立面

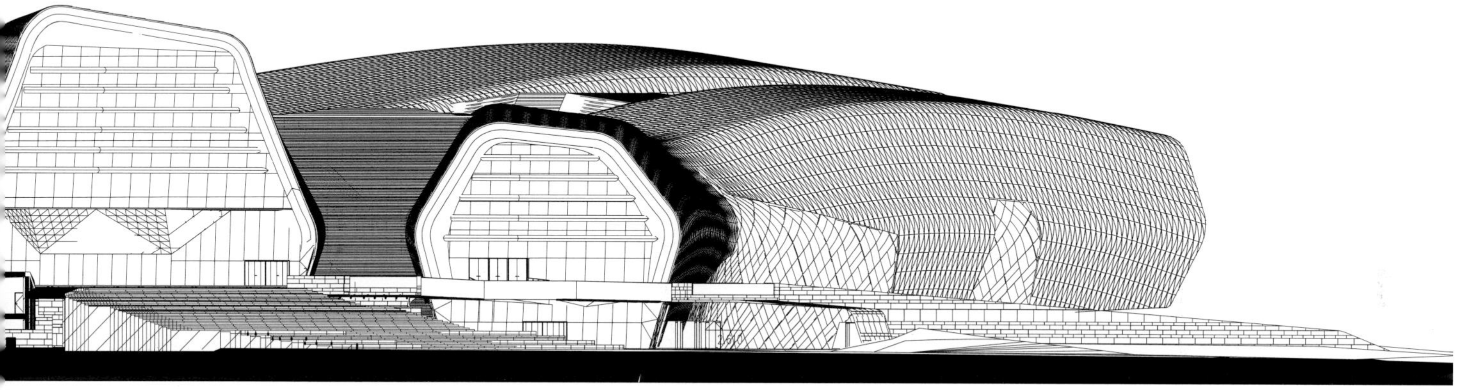

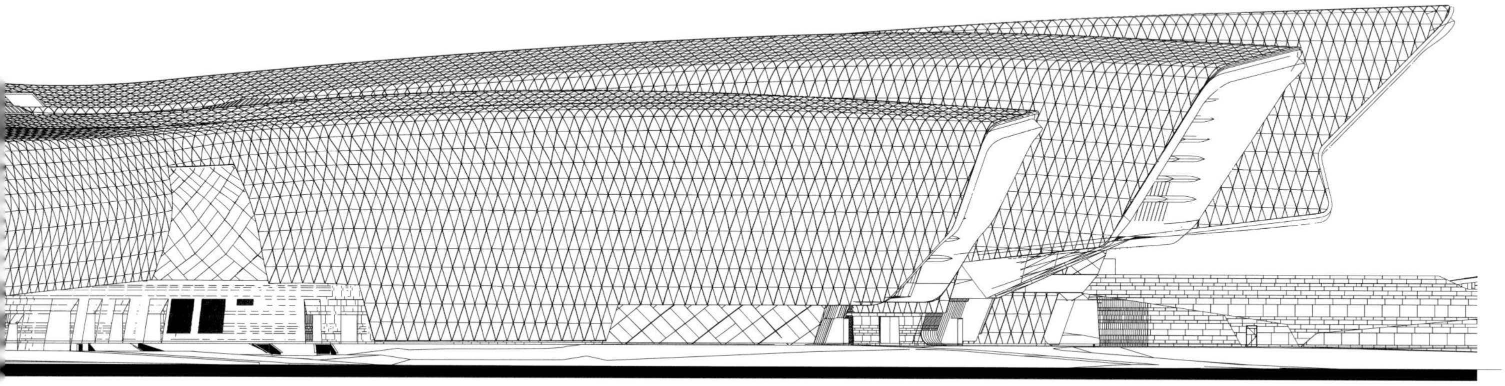

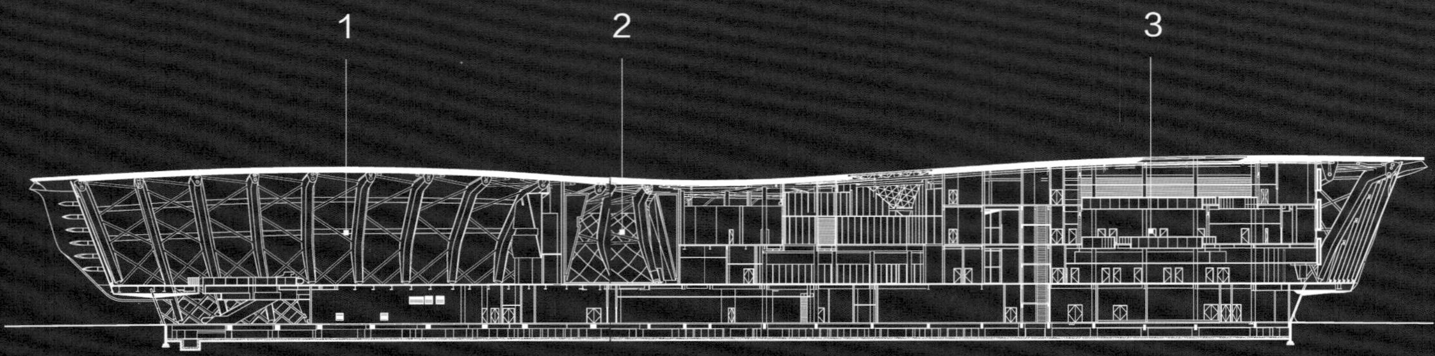

1. 临时展厅　2. 中央大厅西侧厅　3. 办公区共享空间

1. 发现之旅展厅　2. 龙的时代展厅　3. 今日海洋展厅　4. 远古海洋展厅

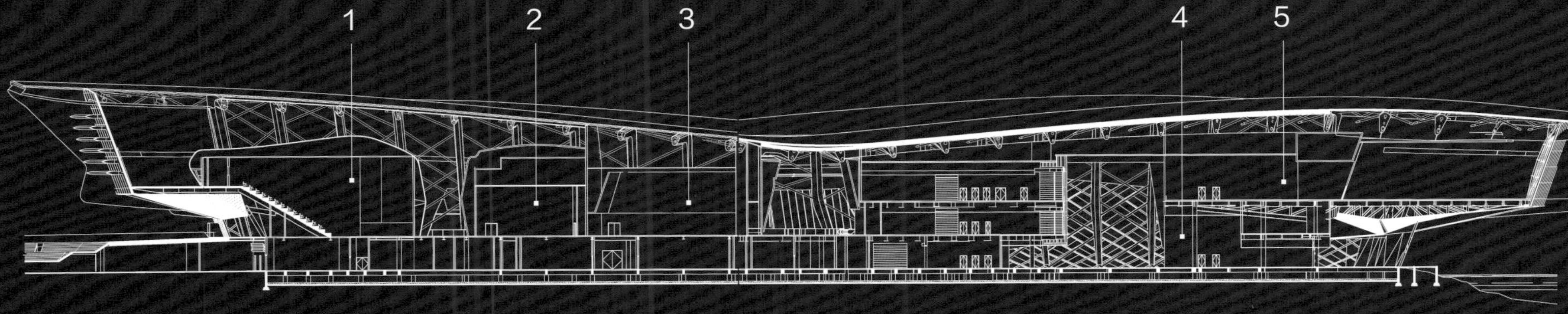

1. 电影院　2. 多功能厅　3. 中华海洋文明第一篇章　4. 海洋文化主题空间　5. 海洋天文

1. 蓝色家园　2. 中华海洋文明第二篇章　3. 从风帆到行轮　4. 穿越极地

剖面（组图）

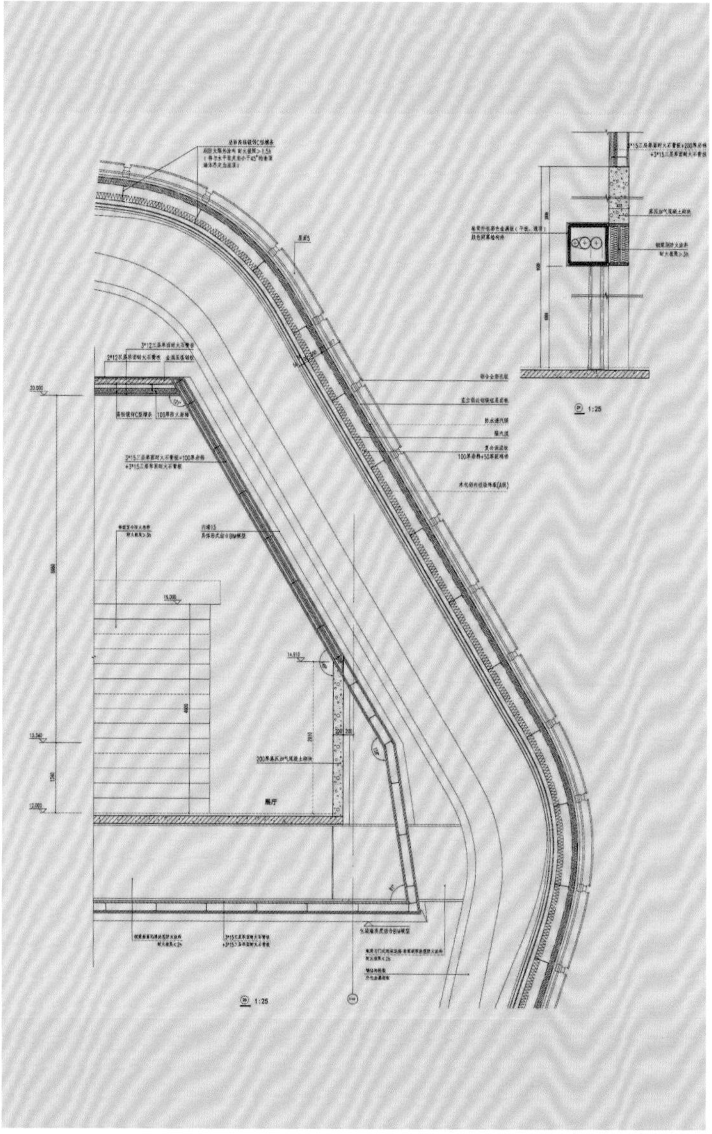

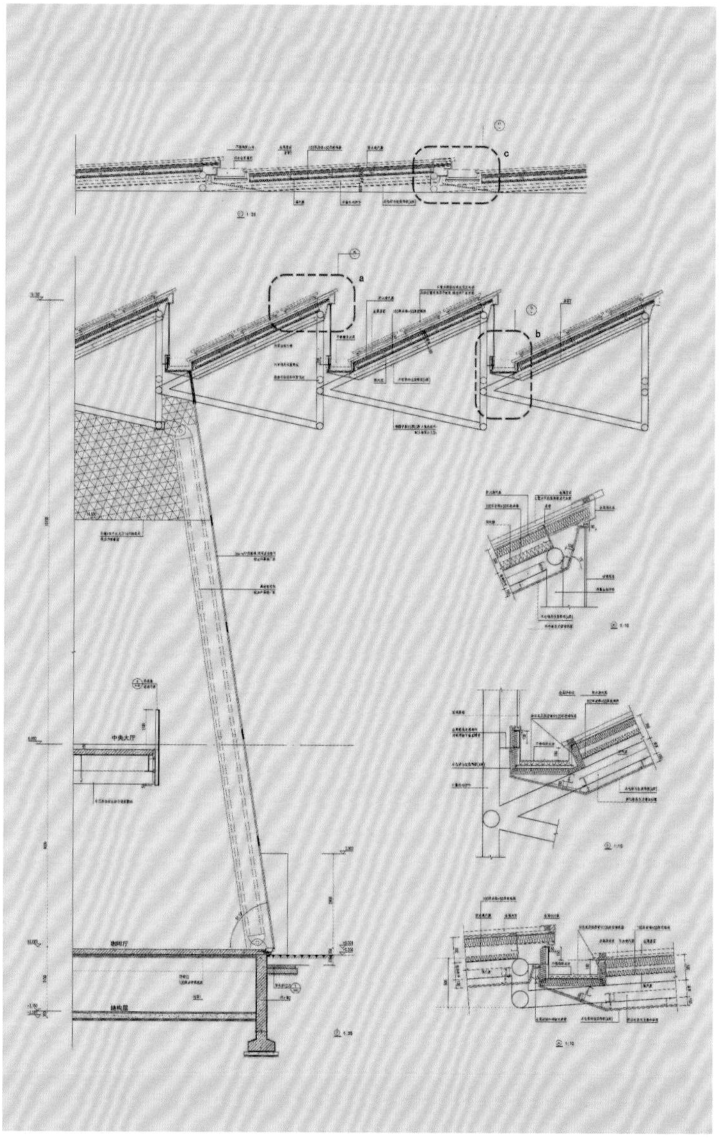

节点详图（组图）

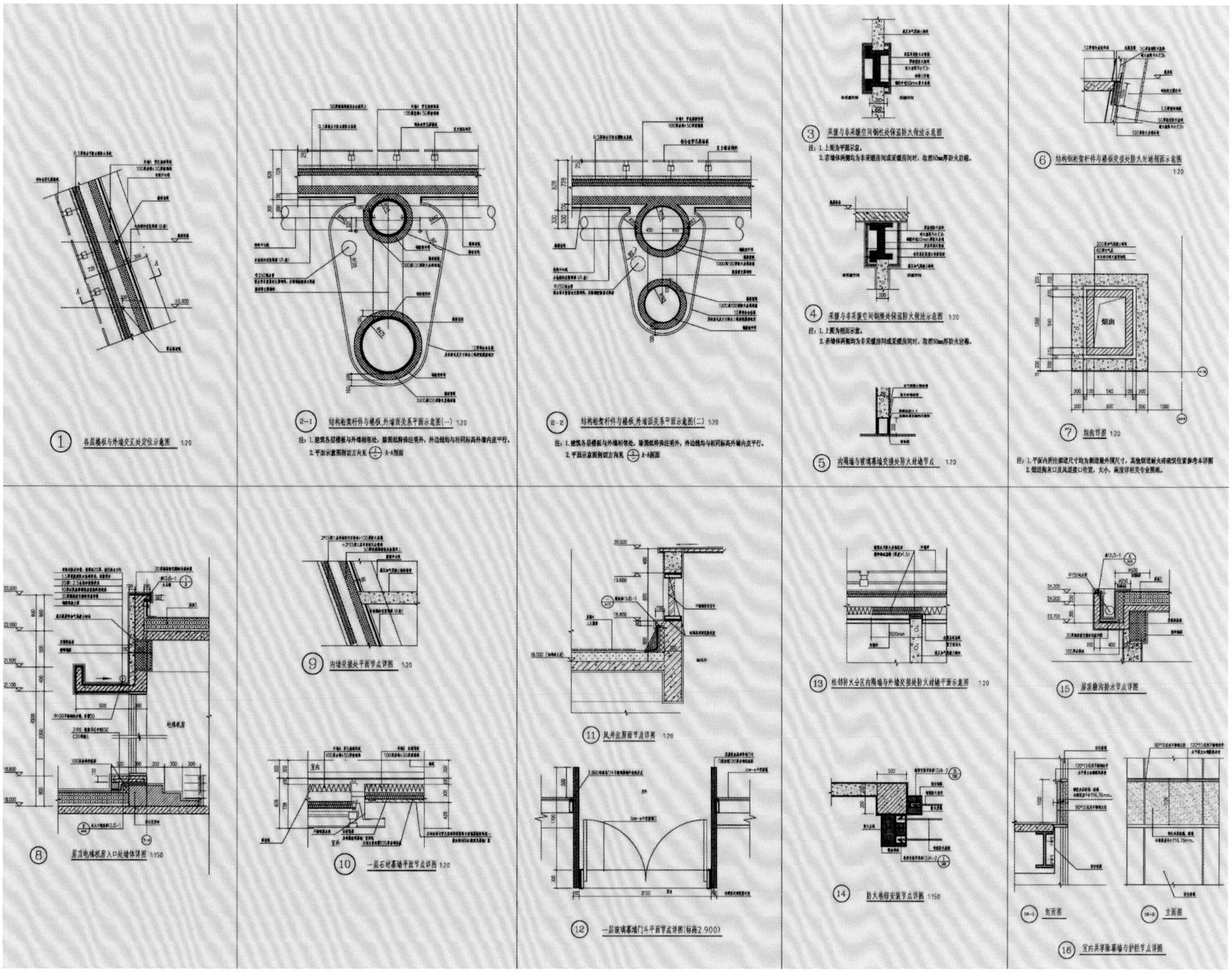

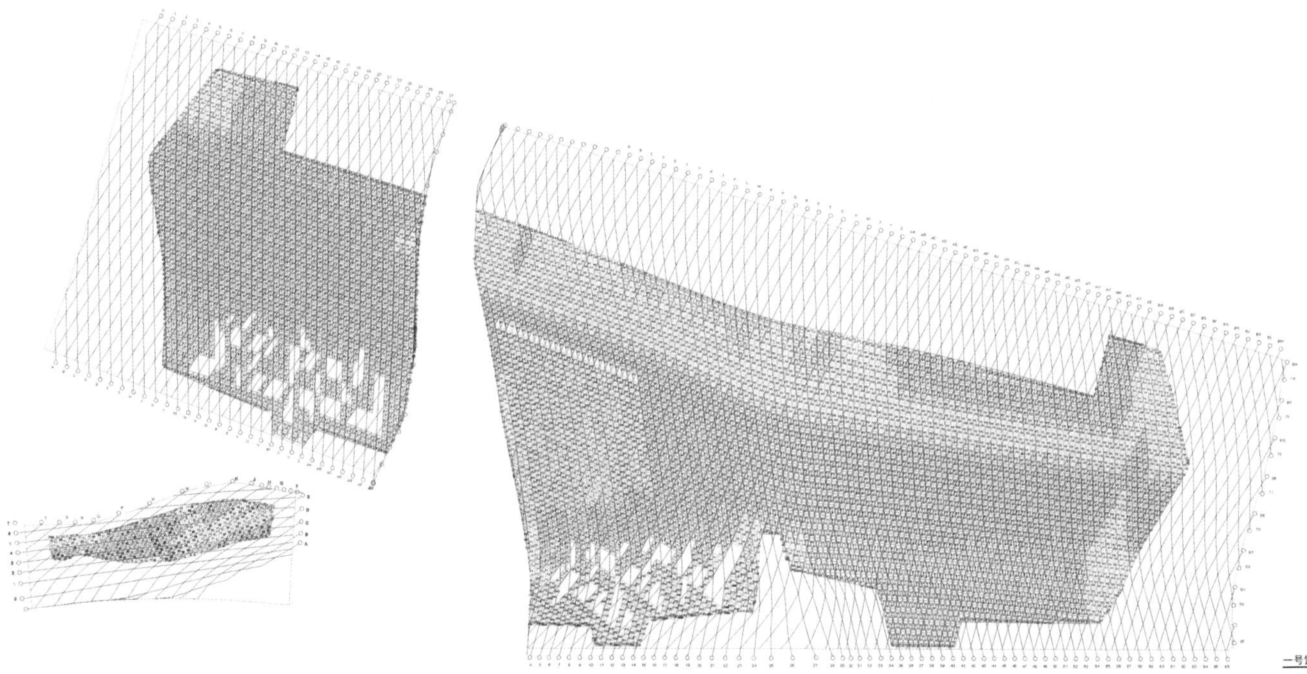

一号馆铝板幕墙展开图

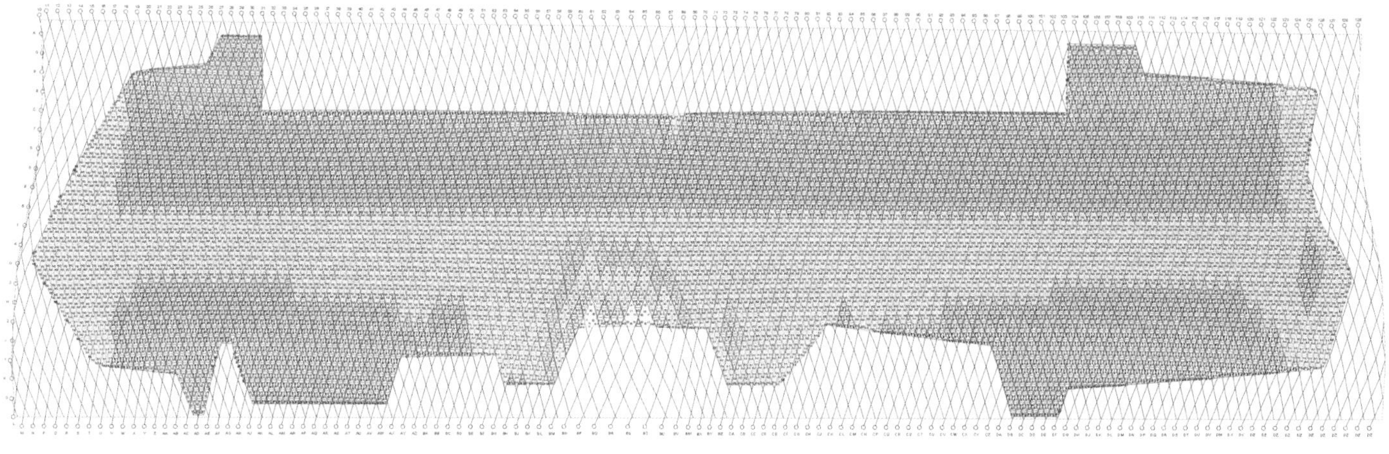

二号馆铝板幕墙展开图

金属幕墙展开示意（组图）

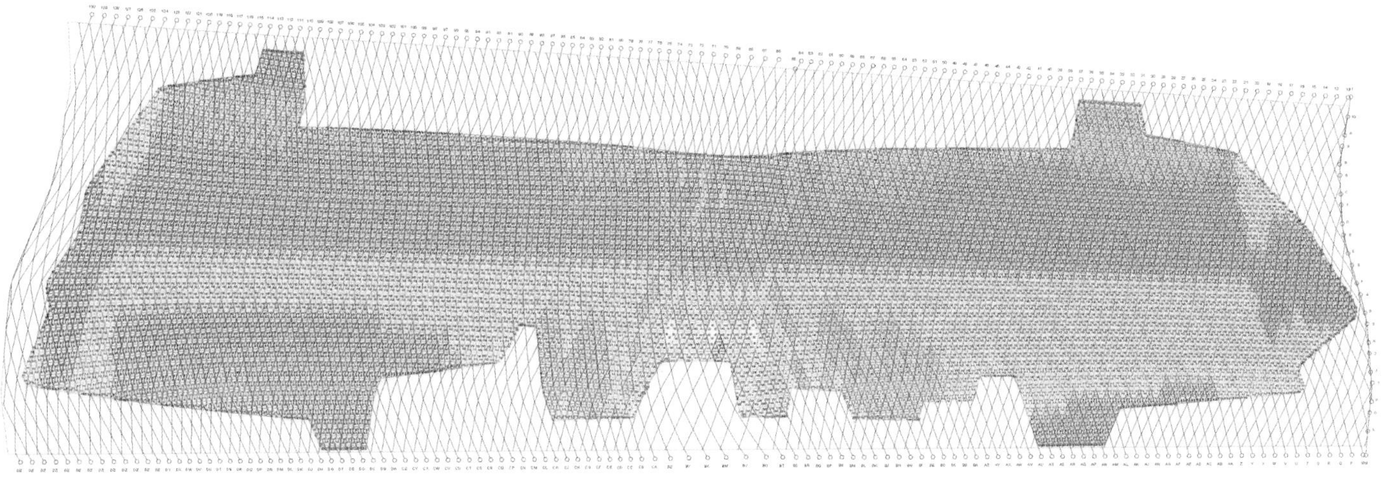

三号馆铝板幕墙展开图

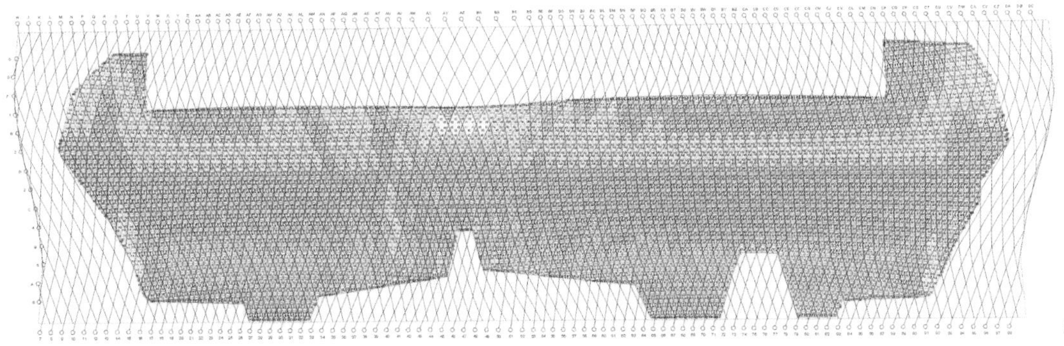

四号馆铝板幕墙展开图

BIM 设计过程图

BIM 设计解决了非线性建筑形体设计的四项技术难点：

- ●建筑形体内外空间的整合；
- ●建筑形体与结构体系的交互设计；
- ●建筑表皮有理化；
- ●建筑内部空间与设备管线的集成。

门式桁架：112 榀

金属幕墙：5.5 万 m^2

玻璃幕墙：1.5 万 m^2

石材幕墙：1.0 万 m^2

视频观赏 1

国家海洋博物馆 BIM 模型

视频观赏 2

国家海洋博物馆全景漫步

视频观赏 3

国家海洋博物馆航拍掠影

五、建筑空间与展陈设计整合的技术支撑

建筑是博物馆建设的基石和载体，如今的博物馆已经成为公众社会生活越来越不可或缺的组成要素，博物馆建筑的合理性、科学性和舒适性更深度地影响着公众博物馆之旅的总体感受。随着时代的变迁，公众除了对展示有更高要求，也更关注博物馆的整体环境与氛围。

从一开始建筑师就结合展品布展高度对空间进行定位，并进行观众观展可视面高度、展具设计高度与建筑状况的匹配设计。这些丰富变化的建筑空间和最恰当的多元展示方式对接融合，为呈现出一个在中国甚至世界都具有自己鲜明个性的杰出海洋博物馆提供了依托。营造富有变化的共享空间，可使观众置身于一个与他人、与展品、与环境、与建筑实时和动态的互动之中。

中央大厅东侧贴临贵宾接待厅、多功能厅和沉浸式全景影院

1. 构建绿色、阳光的中央大厅空间

中央大厅是博物馆交通的主枢纽，大厅北端向东西两侧延伸，与一至四号厅有机贯通，形成有序的立体化的交通空间组织。

中央大厅两侧设有贵宾接待厅、学术报告厅和专卖商店、自助餐厅以及极具特色的沉浸式全景电影院。

中央大厅利用屋面锯齿形的结构构造引入太阳能热水和光伏发电的自然能源利用系统，不仅体现绿色、阳光的交通枢纽，而且彰显中央大厅的庄重大气，富有礼仪感。

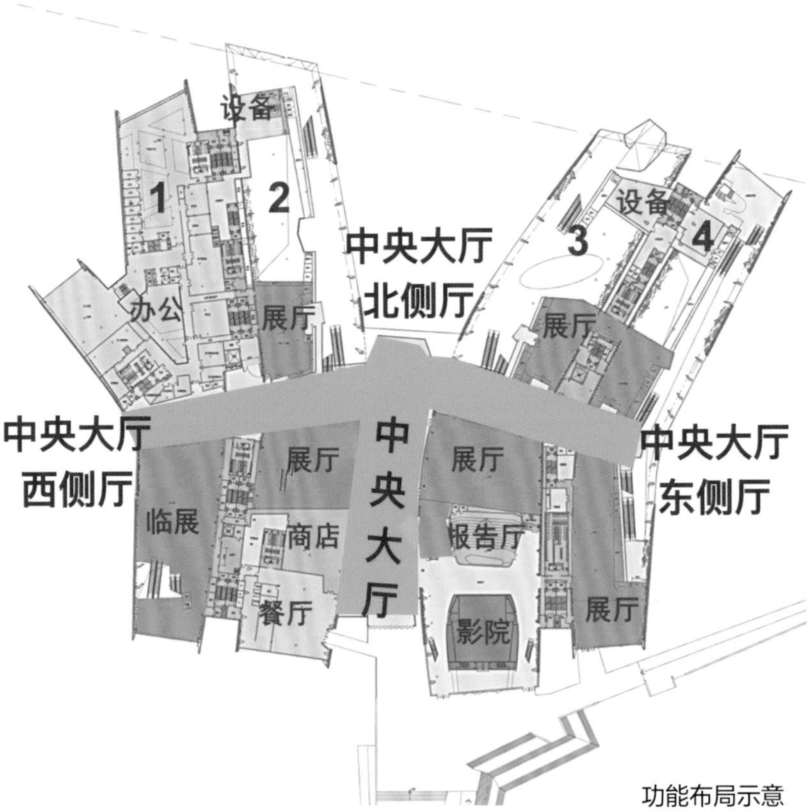

功能布局示意

中央大厅西侧贴临文创纪念品专卖商店和观众自助餐厅

中央大厅

贵宾接待厅

学术报告厅

国家海洋博物馆沉浸式全景影院是最前沿的科技交流平台和标志性文化互动设施，建成后将以全新的视听效果展示海洋自然历史和人文历史，成为集影视观摩、演艺表演、科学交流、娱乐互动等功能于一体的国际化的海洋科普教育影院。观众席由固定坐席区（359 座）、站立观影区（40 席）组成。

该影院可以满足影视观摩、海洋探秘、科学互动、演艺表演等需求，通过空间移动转换、声道立体音响、多用途的无缝切换，打造梦幻式的舞台，给予观众沉浸式的视听体验，使观众有一种置身于虚拟海洋世界之中的感觉。

沉浸式影院配置完善的全景幕多媒体系统，可满足全景五幕、3D 半环幕、标准 2D/3D 综合演艺等多种使用需求，该系统主要由以下各系统集成组成：机械切换系统、多媒体投影系统、LED 屏系统、扩声系统。

①全景五幕模式由主幕、左侧幕、右侧幕、天幕（LED）、地幕（LED）共同构成。

② 2D/3D 标准放映模式是由主幕独立实现的最基础的模式。

③ 3D 半环幕模式由主幕、左侧幕、右侧幕共同构成。

④在综合演艺模式下，主幕可移动至舞台后墙，空出舞台表演区，满足了小型演出、会议论坛的使用需求。

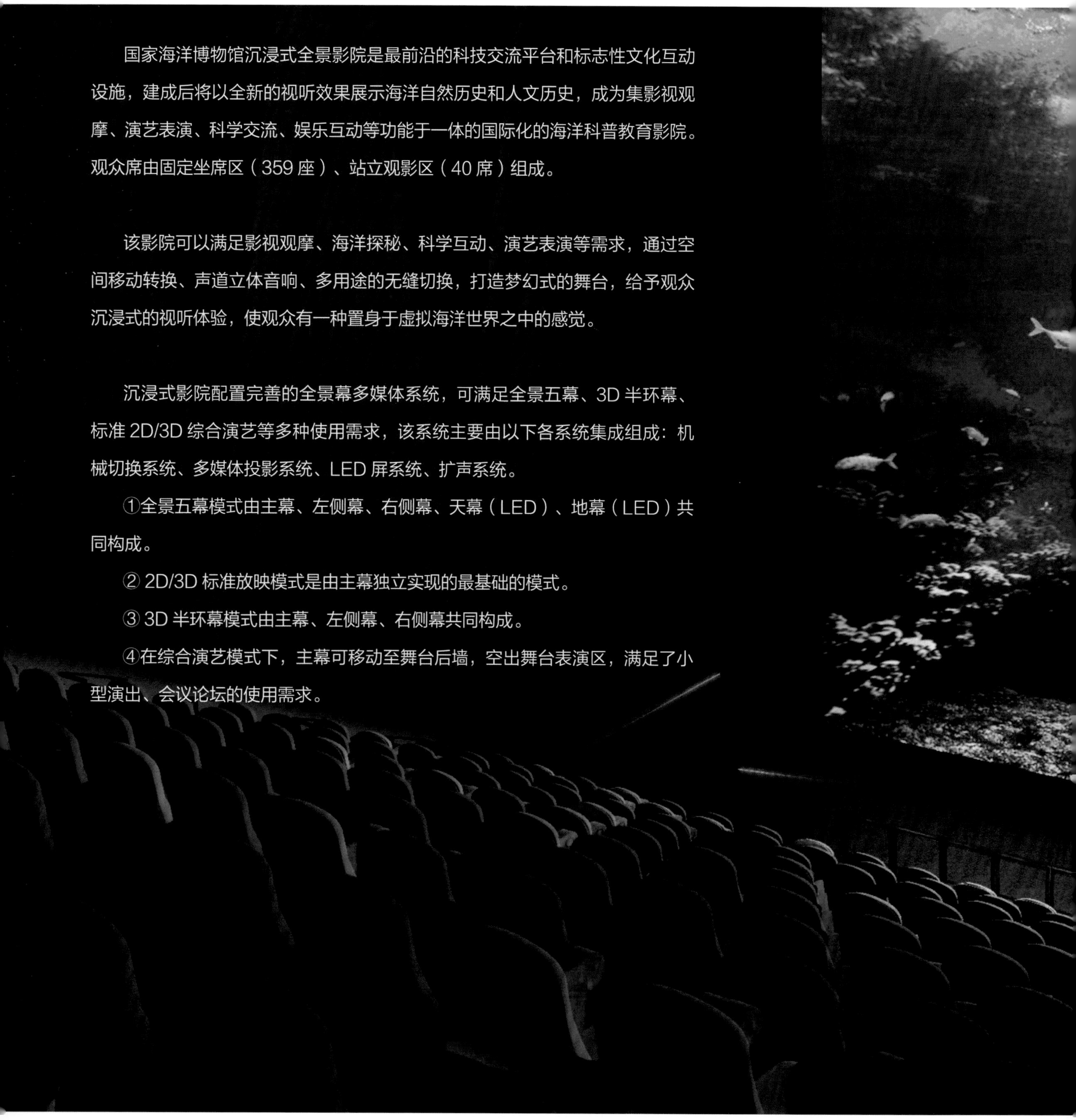

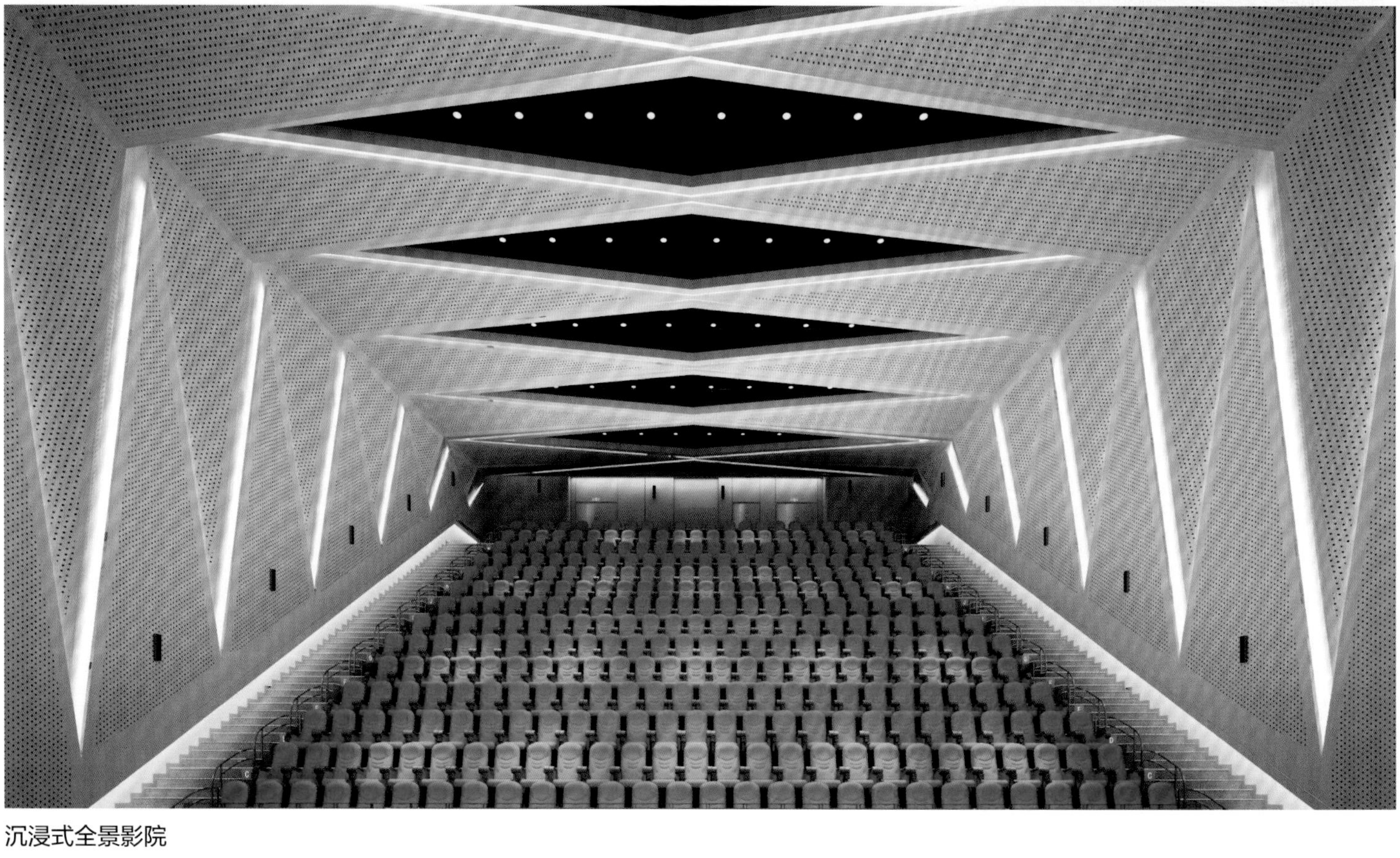

沉浸式全景影院

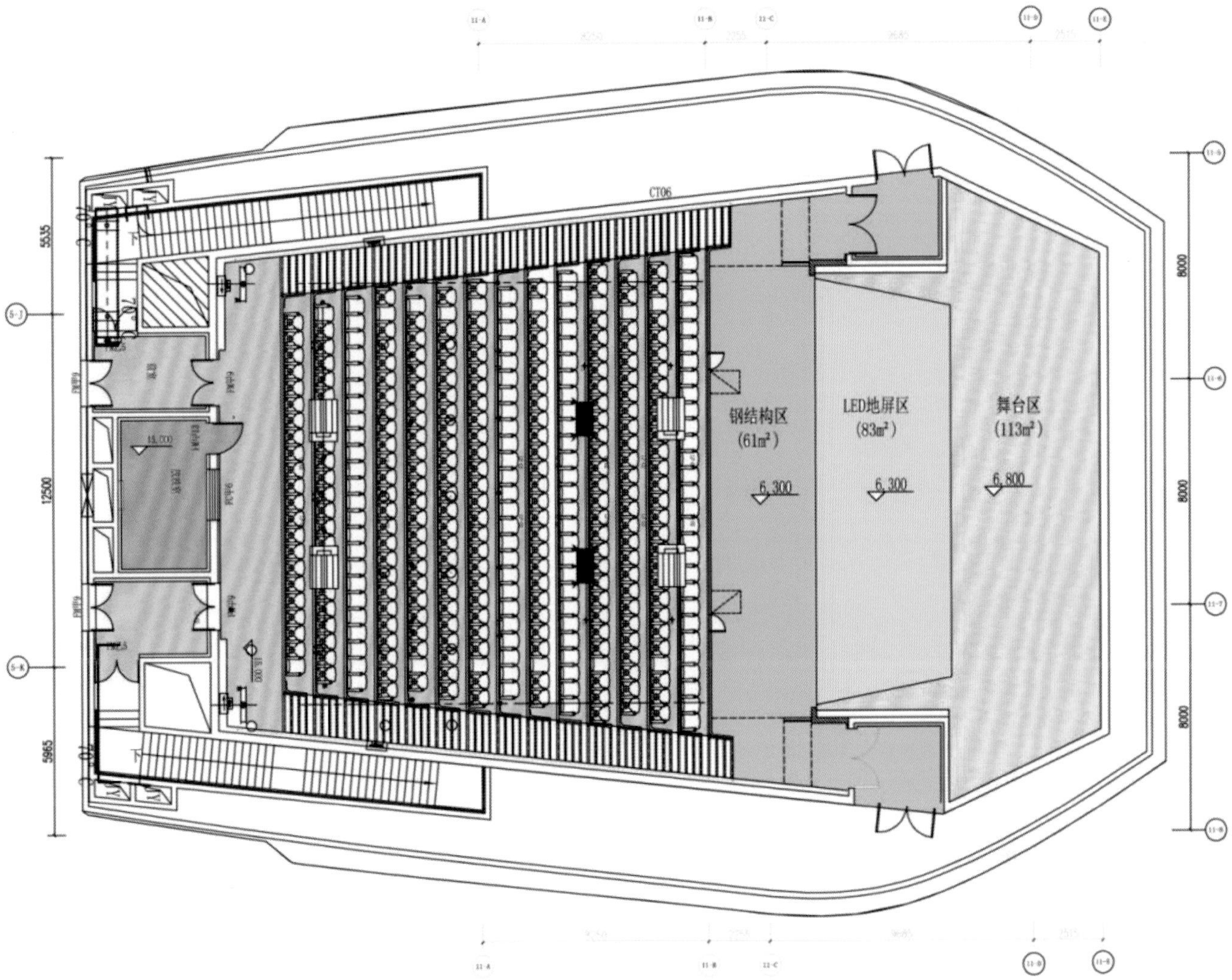

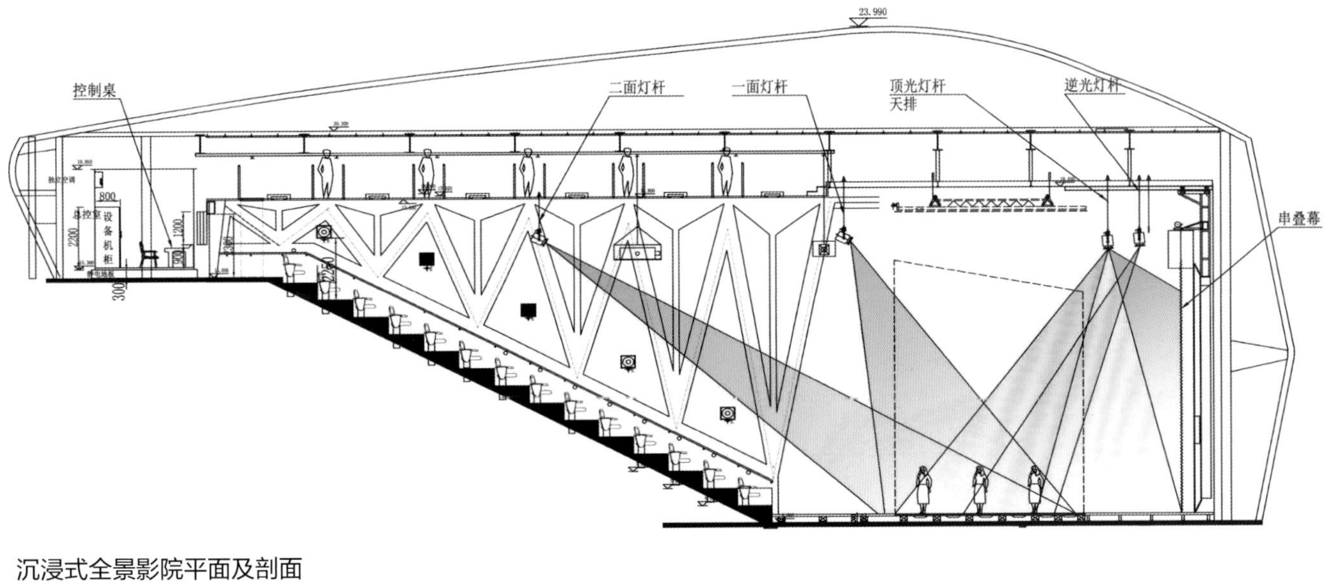

沉浸式全景影院平面及剖面

沉浸式观影模式示意（组图）

青春心向党 建功新时代
纪念五四运动100周年

纪念品商店

观众餐厅

中央大厅西侧厅

中央大厅东侧厅

临时展厅

2. 建筑空间有机整合与展陈设计的融合

将建筑设计为多元化展陈主题布展空间，营造收放自如、富于张力的艺术展示环境，是建筑空间设计的创新点。结构设计通过突破大跨度无柱空间、跨层跃层超高空间、超大尺度悬挑空间等技术难题，对空间有机整合，充分满足多元化空间布展及以“公众为本” 的观展需求，彰显展厅场域的整体感和连续性，为公众观展带来了开阔的空间视野享受，场景的震撼力以及视觉的冲击力。

展厅内景（组图）

“今日海洋”展厅建筑与展陈的结合

解析海洋
深入海洋
大洋中脊
大洋盆地
大陆边缘
海洋地理构成
海与洋
俯瞰海洋
全球海陆分布

“龙的时代”展厅建筑与展陈的结合

MOVEMENT
SKY
捕食 PREDATION
防御 DEFENSE

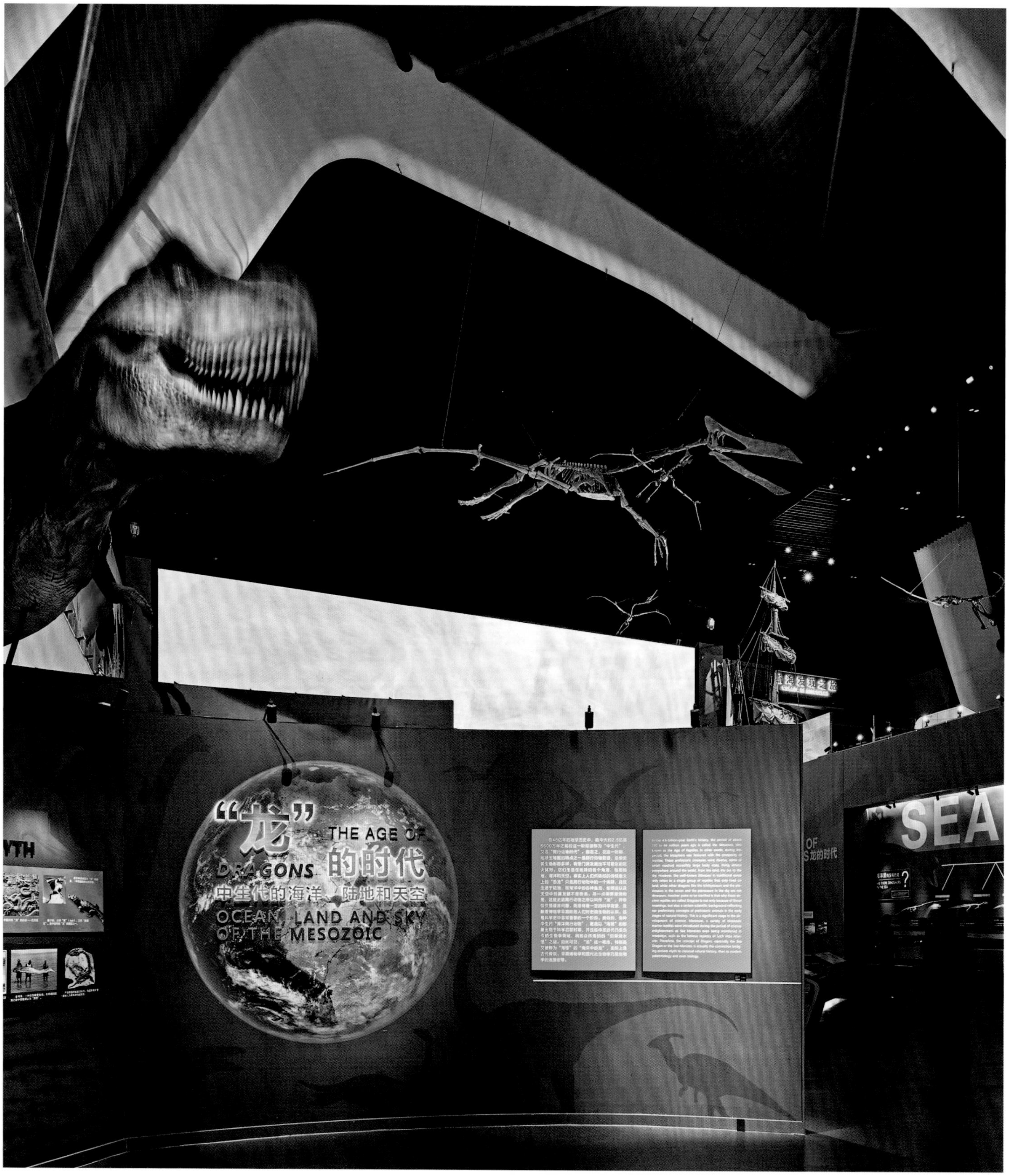

“龙”的时代
THE AGE OF DRAGONS
中生代的海洋、陆地和天空
OCEAN, LAND AND SKY OF THE MESOZOIC
SEA

“龙的时代”展厅（组图）

运动 MOVEMENT
捕食 PREDATION

THE AGE OF
DRAGONS龙的时代
SEA SK

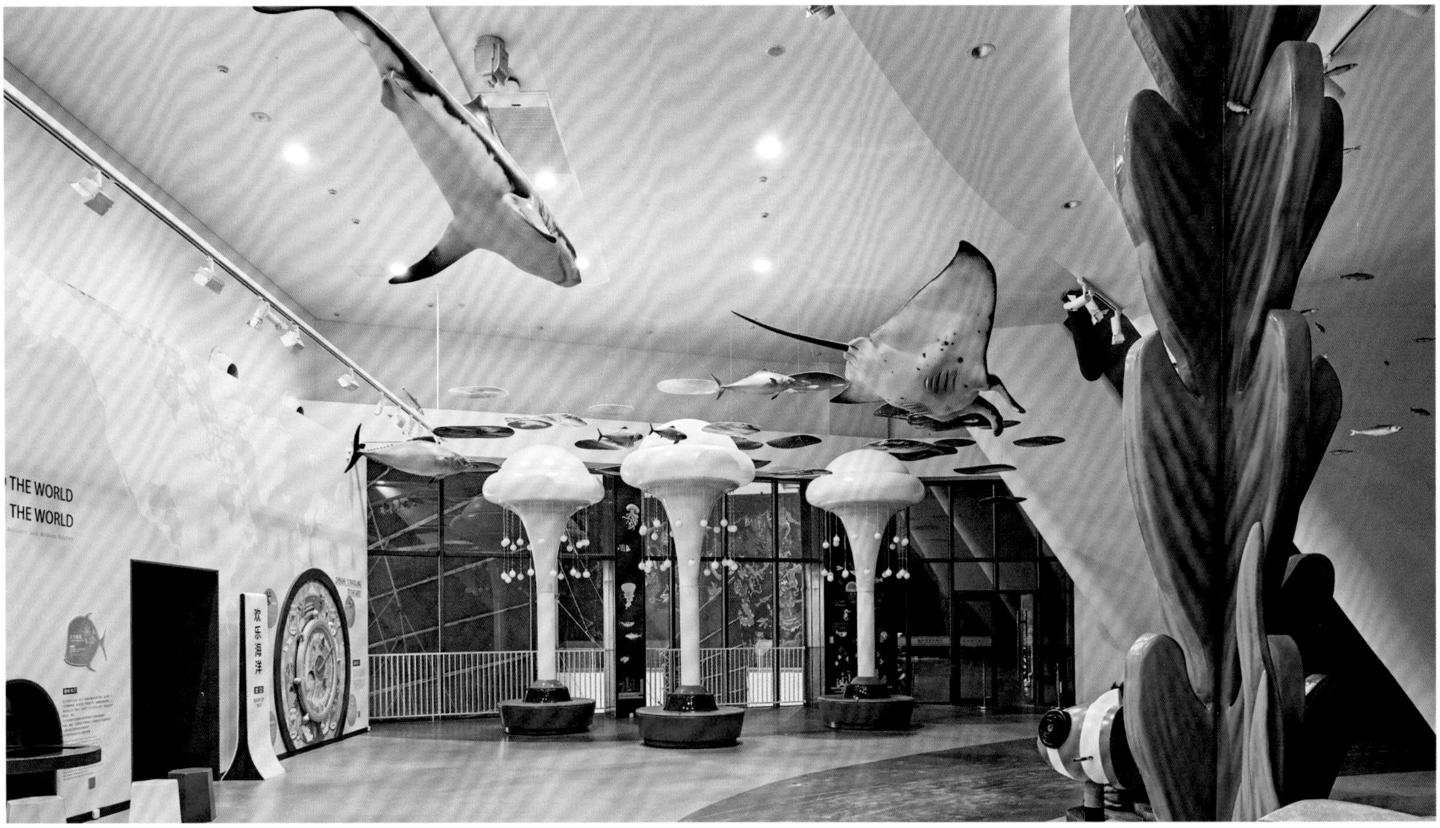

“筑梦极地”展厅（组图）

趣味海洋

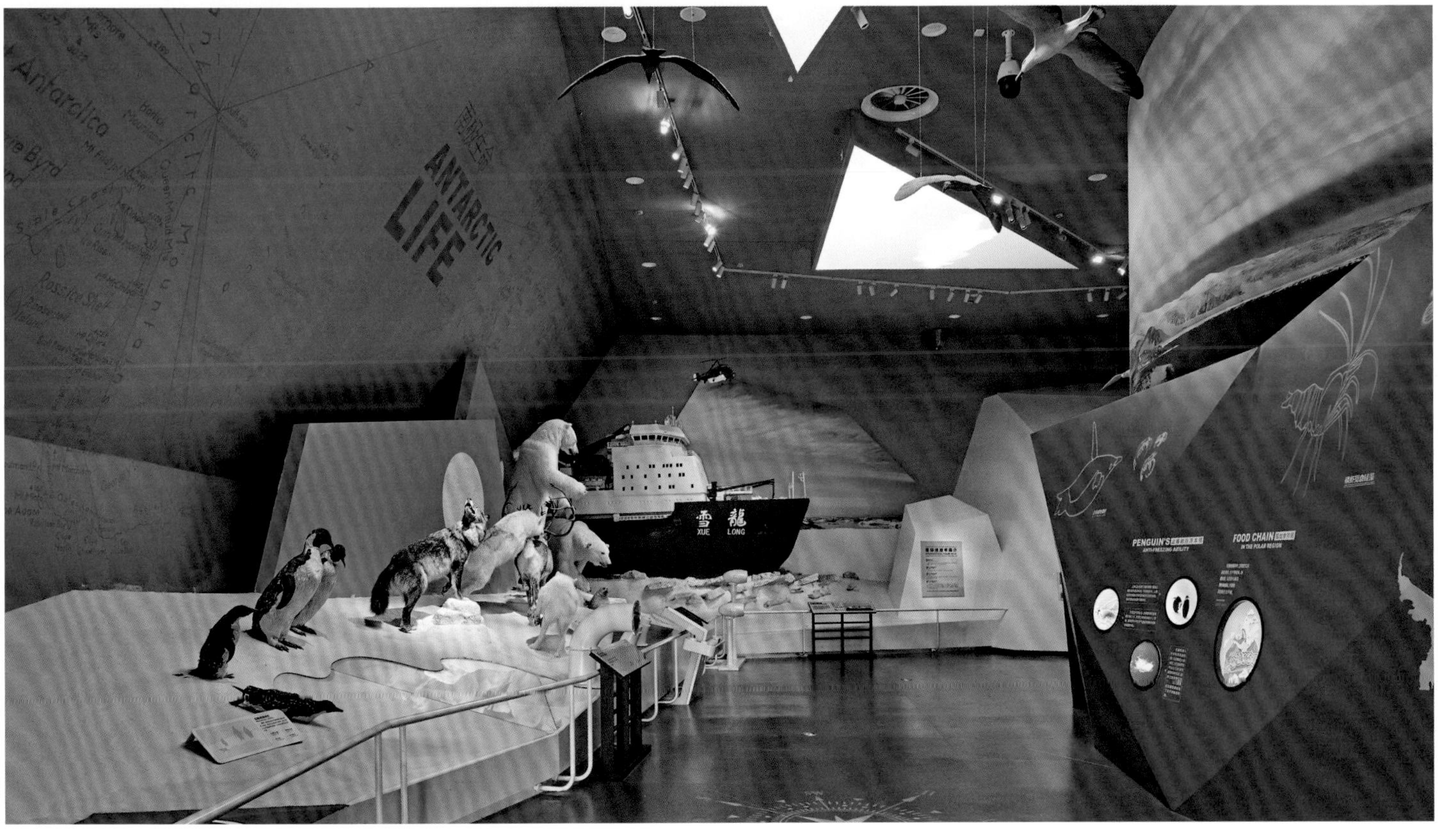
ANTARCTIC LIFE
雪龍
XUE LONG
PENGUIN'S
ANTI-FREEZING ABILITY
FOOD CHAIN
IN THE POLAR REGION

海洋实验课堂建筑与功能的结合

EXPERIMENT CLASS
国家海洋博物馆
NATIONAL MARITIME MUSEUM

3. 跨层跃层超高空间

为满足特殊超高超大展品的布展需求，设计超高空间结构，如 3 号厅宋代福船展位高 28 米。

欢乐海洋

海洋天文厅
海洋天文

福船在超高建筑中的展示（组图）

4. 内外互动的光环境

为了实现馆内自然采光并打造观景的可视性，设计在大体量封闭式展厅的公共交通空间外侧开发临海开放的视觉空间界面，通过几处内外光环境互动的通高交通空间，为观众提供可在馆内尽享大海辽阔海景的条件。

内外互动的光环境（组图）

内外互动的光环境 2（组图）

临北侧开放空间的咖啡厅（组图）

3 号厅交通空间

5. 结构超大悬挑空间

大尺度悬挑结构不仅在海湾中彰显了建筑强烈的视觉冲击力，而且为大型实物展项营造了特有的的外展陈列展位。

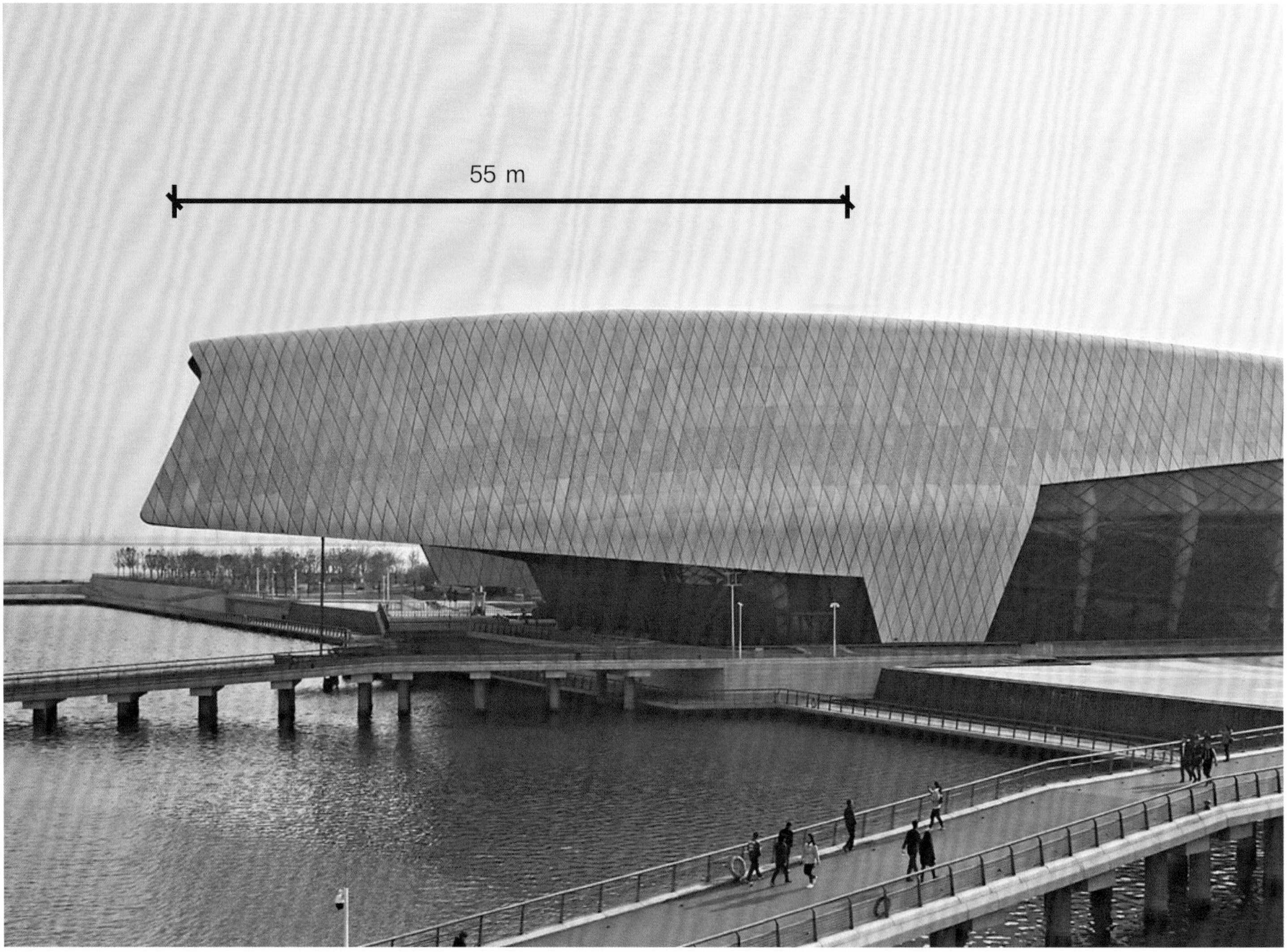

55 m 长度的挑檐示意

夜色下的挑檐

六、夜景景观文化

伴随社会经济的发展，人们对夜景文化的体验已成为生活中的一个不可获缺的重要需求，在兼顾功能性和体验性的前提下，关注提升中新天津生态城地域的夜景文化品质，保障“文旅融合”开发的整体环境，是本设计的重要组成篇章。

夕阳下的国家海洋博物馆（组图）

夜幕降临后的海洋博物馆

华灯初上

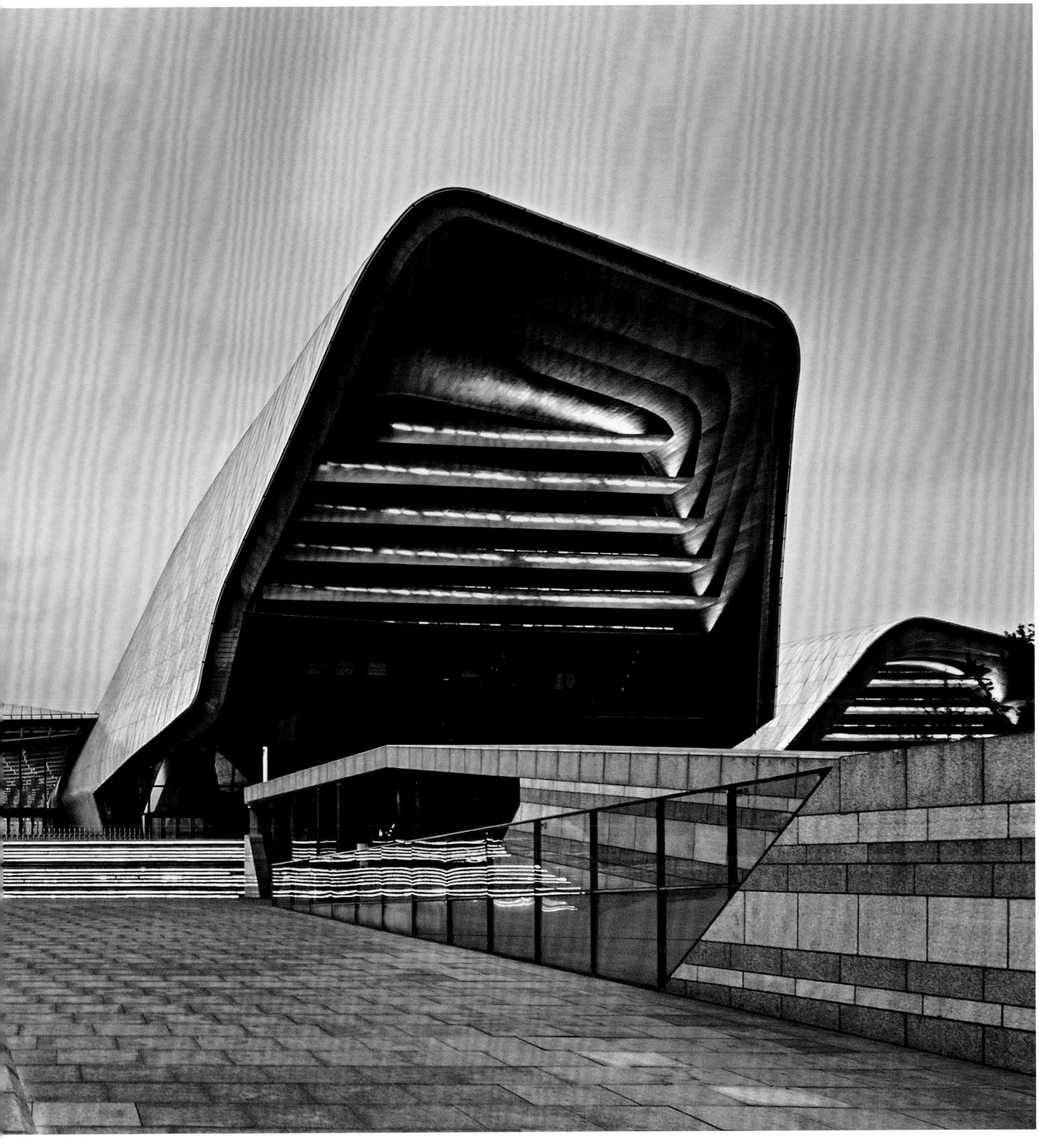

建筑主体的灯光在夜空中

建筑主体的灯光在夜空中 2

第三章　专业设计

Chapter III Professional Design

国家海洋博物馆的设计和建造彰显了各相关专业的交织是一个庞大和复杂的工程。不同专业间严谨有序的协调与融合，是确保展馆作为一个高完成度整体完美呈现的关键。正是建筑设计、装饰设计、结构设计、景观设计等专业与建筑布展设计的无缝对接，体现了各专业人员对项目总体设计的把控能力，从而成就了非凡的观感与内涵。

The design and construction of the National Maritime Museum of China demonstrate that the interweaving of various specialties concerned is a huge and complex project. Strict and orderly coordination and integration among different specialties is the key to ensure the perfect presentation of the museum as a whole with a high degree of completion. It is the seamless integration among architectural design, decoration design, structural design, landscape design and architectural layout design that embodies the ability of professionals of different specialities to control the overall design of the project, thus accomplishing an extraordinary visual impact and presenting remarkable contents.

一、装修设计

在博物馆整体建筑空间中，设计团队以全程化服务统筹非标准化设计、延续化设计、一体化设计、精细化设计，圆满完成了国家海洋博物馆的室内装修专业设计。

非标准化设计。如果说建筑体宛如飞跃入海的鱼群，那么 112 榀主体钢桁架就如同“鱼骨”一样，支撑起各个展馆的“鱼腹”空间。这些钢桁架的技术数据各不相同，形态各有差异，没有一个标准化的断面。千变万化的钢桁架造型丰富，给室内装修的设计与施工增加了难度，为此，装修设计选用 GRG 材料对结构桁架进行包设。GRG，即预筑式玻璃纤维石膏板，其流体预筑式的生产方式使定制任意造型产品成为可能，设计者能根据设计图转化生成加工制造图。另外，GRG 质量轻，强度高，既能通过微孔结构吸声降噪，还能通过呼吸效应调节室内相对湿度，是营造室内环境的理想材料。采用 GRG 包设的“鱼骨”，每一根都没有明显接缝印记，彰显了形态各异骨架的完整形象。

延续化设计。建筑的室内与室外一体两面，密不可分。室内设计是在建筑方案设计构筑的基础上进行的延续性二次创作，是深化设计，其将建筑外檐的元素延续至室内空间，室内望向室外，室外窥至室内，这种统一中求变化的室内设计旨在创造更加和谐优美的博物馆空间环境。

一体化设计。室内装修设计作为建筑设计的一部分，必须与各专业进行一体化设计，在全流程的设计中掌控好每一个阶段的细节，表达出最优的设计成果，并体现出一种整体统筹的能力。尤其是机电专业的终端设备来说，它们需要和室内造型协调统一，以展现更加完美、高品质的建筑空间。例如：无论是通高展厅的 GRG 包装体上，还是木纹铝板的包装墙面上，装修设计时都做了设备槽，在满足规范的前提下，将机电专业的终端点位集中有序地布置在设备槽中，以实现集成度高、色调统一、隐蔽性强、和谐美观、检修方便。

精细化设计。细节决定成败，一个优秀的室内装修离不开对细节设计的思考。即使是石材阳角，也做了磨圆倒角处理，对这些细节的推敲与思考，体现了以人为本的设计宗旨和“工匠精神”，提高了整个工程的品质及效率。

全程化服务。为了让作品呈现出最好的效果，秉承对工程负责的态度，设计师全过程驻场，对每个空间、每个细节进行设计配合及监督，及时解决施工方的疑点和技术难题，从而完成一座创新的国家海洋博物馆的装修设计。

1. 非标准化设计

用 GRG 将门式桁架进行包设，使其宛如“鱼骨”。每一根“鱼骨” 都形态各异，且每根都呈现为没有任何接缝的一个整体。

请勿抛物
NO THROWING

2. 延续化设计

室内设计和建筑设计是相辅相成、密不可分的。室内设计依附于建筑设计，并在其基础上进行深化设计，目标一致并相互渗透。

中国海洋博物馆
Maritime Museum of China
中国海洋博物馆
Maritime Museum of China

3. 一体化设计

GRG 包装体及木纹铝板的包装墙面上都设计了设备槽，集成度高、隐蔽性强、和谐美观、检修方便。

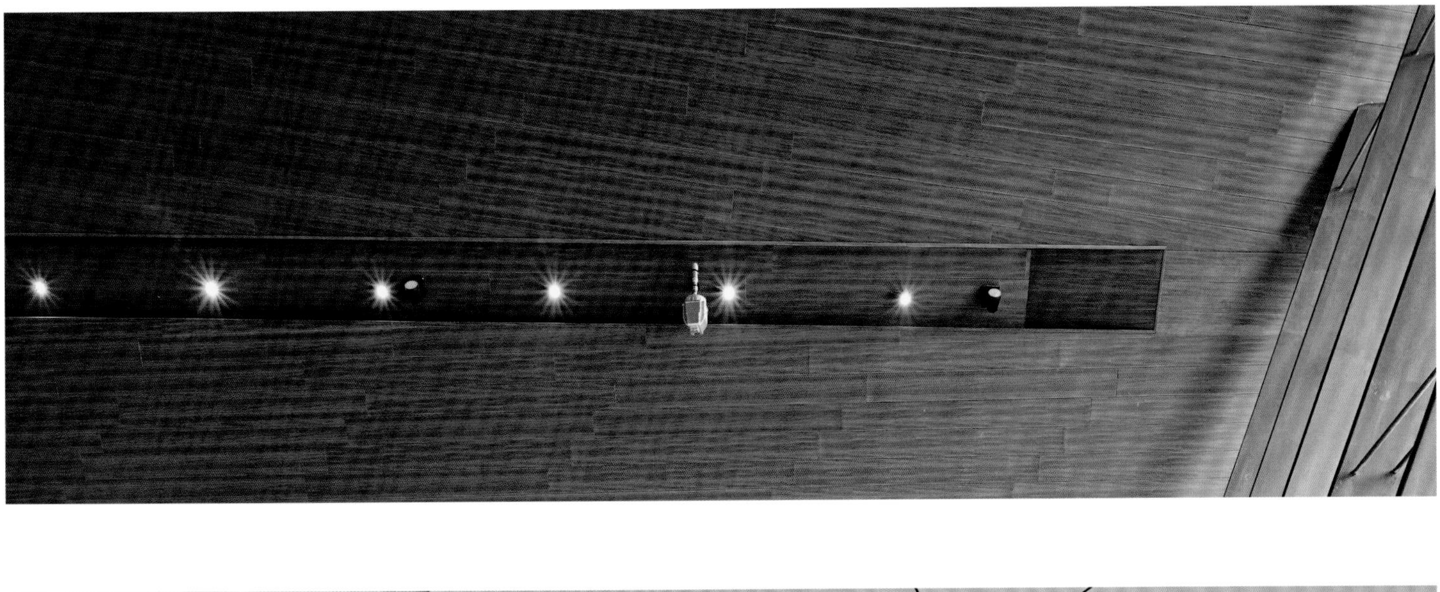

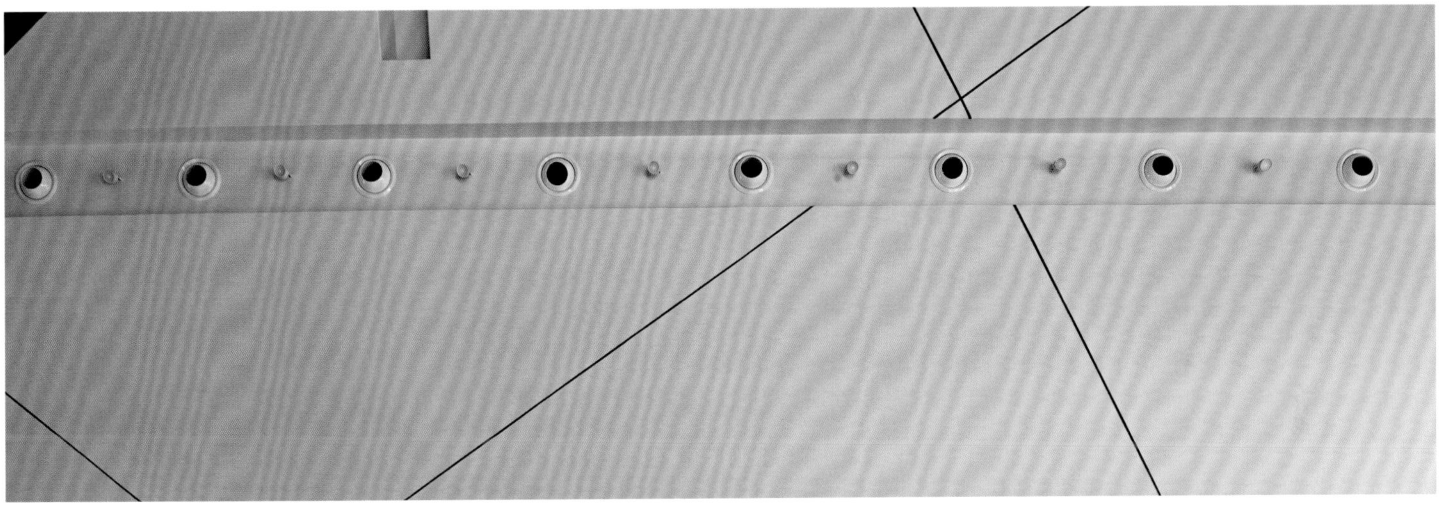

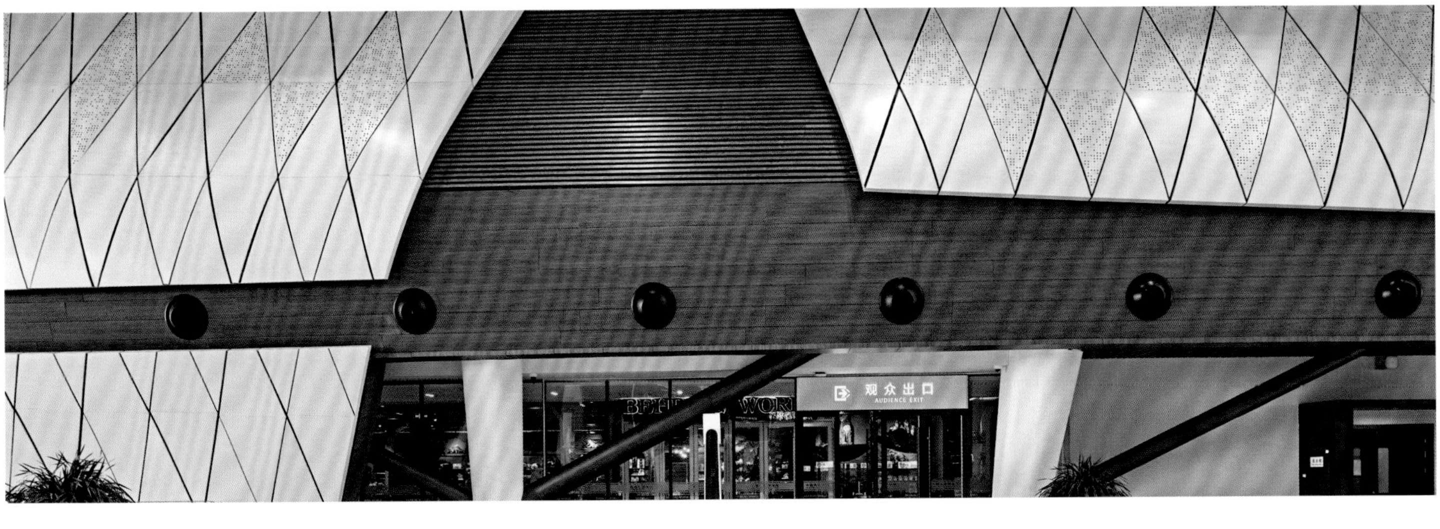
观众出口
AUDIENCE EXIT

4. 精细化设计

公共空间的石材墙面与 GRG 墙面的阳角都做了圆角处理，体现了人性化设计。

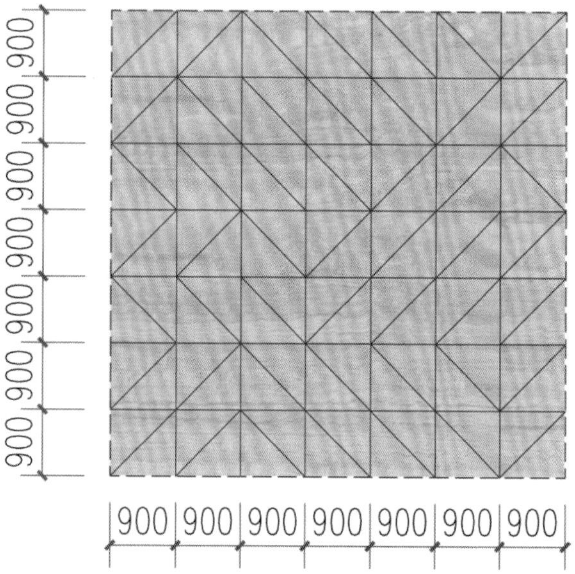

在方案设计阶段，墙面采用三角形石材拼装，地面材料采用洞石。洞石质地软，不适合在人流量大的空间使用，三角形石材则不利于成品保护，后期维护困难。

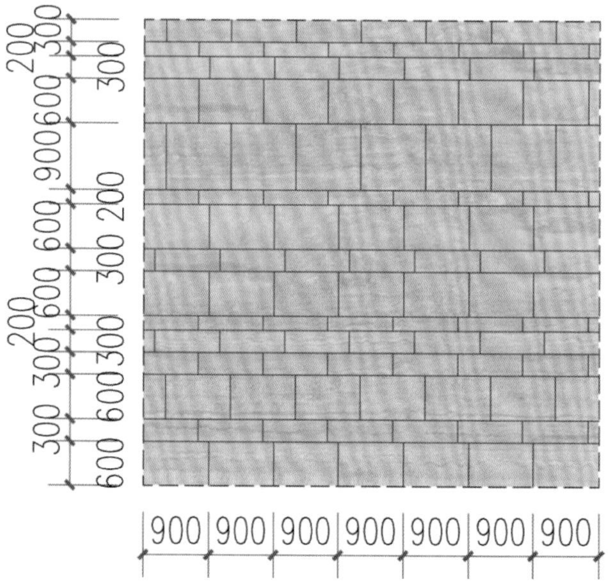

最终公共空间的地面采用质地较硬的奥特曼大理石，墙面石材则由三角形石材变更为不同高度的矩形石材。

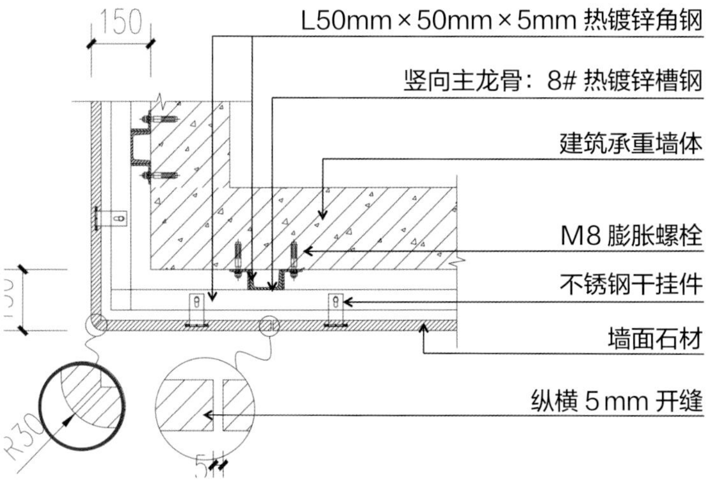
150
L50mm×50mm×5mm 热镀锌角钢
竖向主龙骨：8# 热镀锌槽钢
建筑承重墙体
M8 膨胀螺栓
不锈钢干挂件
墙面石材
纵横 5mm 开缝
R30

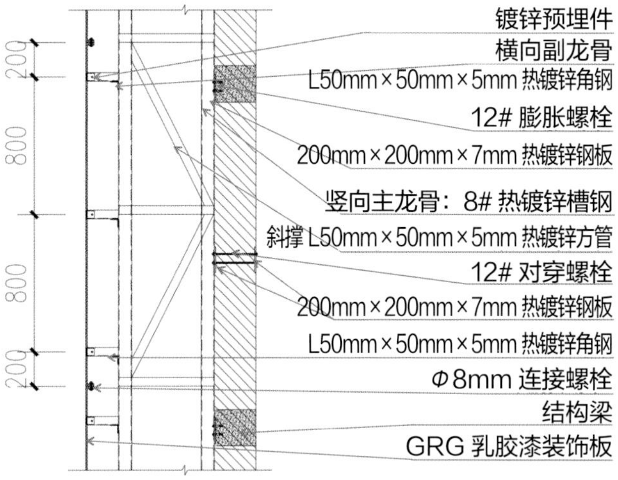
200
800
800
200
镀锌预埋件
横向副龙骨
L50mm×50mm×5mm 热镀锌角钢
12# 膨胀螺栓
200mm×200mm×7mm 热镀锌钢板
竖向主龙骨：8# 热镀锌槽钢
斜撑 L50mm×50mm×5mm 热镀锌方管
12# 对穿螺栓
200mm×200mm×7mm 热镀锌钢板
L50mm×50mm×5mm 热镀锌角钢
Φ8mm 连接螺栓
结构梁
GRG 乳胶漆装饰板

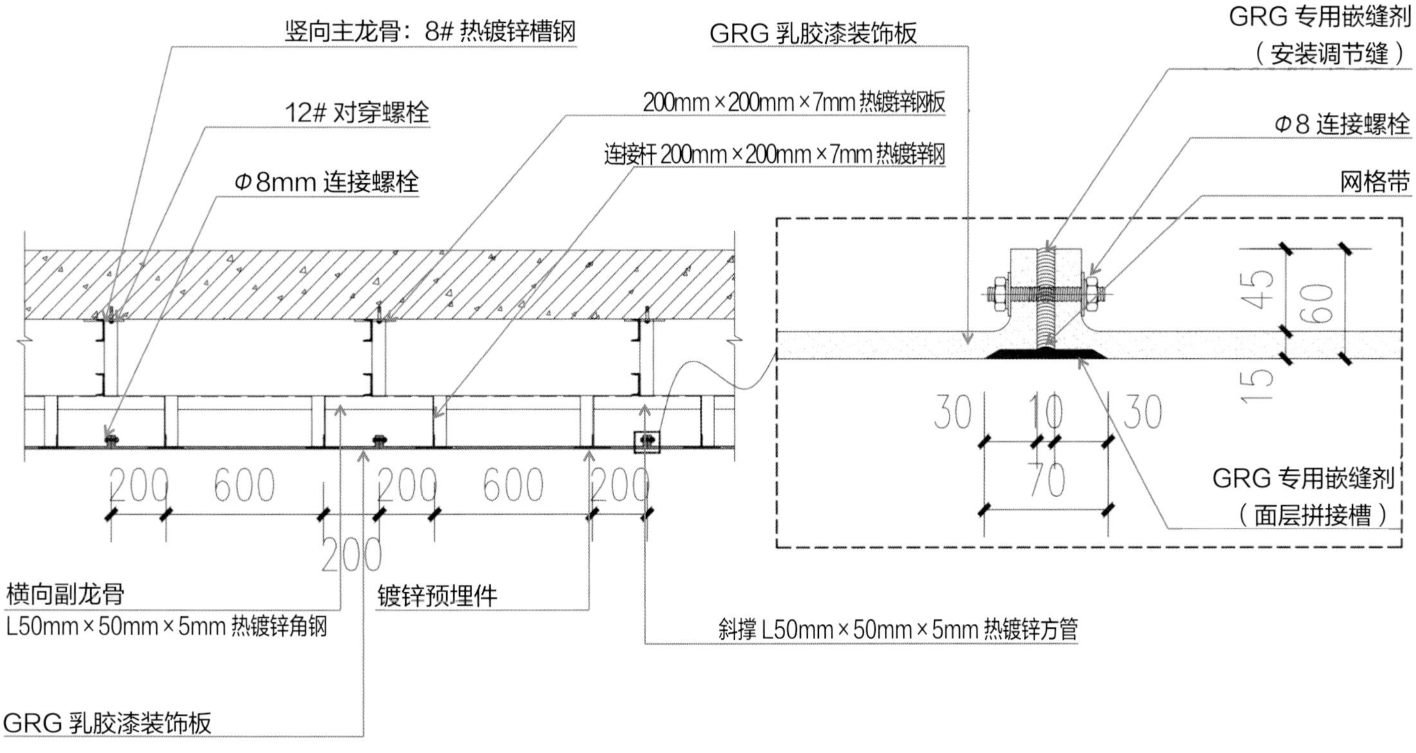
竖向主龙骨：8# 热镀锌槽钢
12# 对穿螺栓
Φ8mm 连接螺栓
GRG 乳胶漆装饰板
200mm×200mm×7mm 热镀锌钢板
连接杆 200mm×200mm×7mm 热镀锌钢
GRG 专用嵌缝剂
（安装调节缝）
Φ8 连接螺栓
网格带
45
60
15
30
10
30
70
GRG 专用嵌缝剂
（面层拼接槽）
200
600
200
600
200
200
横向副龙骨
L50mm×50mm×5mm 热镀锌角钢
镀锌预埋件
斜撑 L50mm×50mm×5mm 热镀锌方管
GRG 乳胶漆装饰板

编号 1（门斗内侧）
设备槽尺寸（宽 × 高）：
150mm×150mm

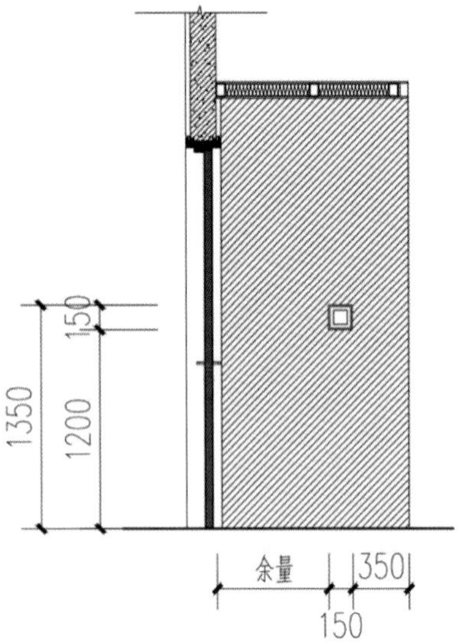

编号 2（门斗内侧）
设备槽尺寸（宽 × 高）：
150mm×280mm

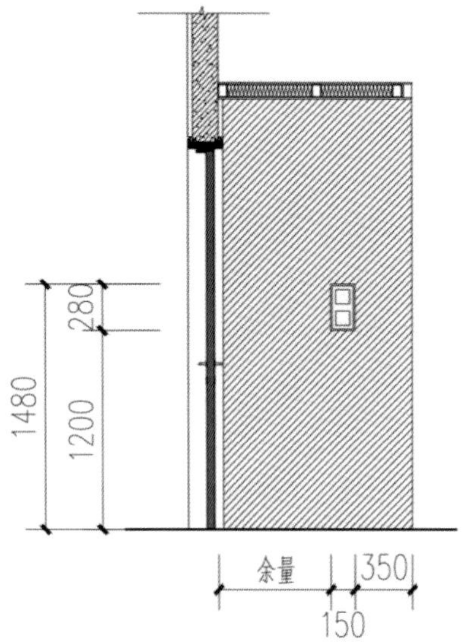

编号 3（门斗内侧）
设备槽尺寸（宽 × 高）：
150mm×450mm

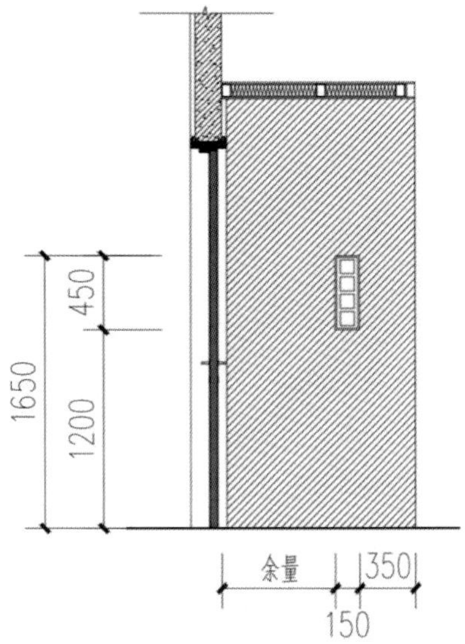

编号 4（门斗外侧）
设备槽尺寸（宽 × 高）：
150mm×150mm
150mm×200mm

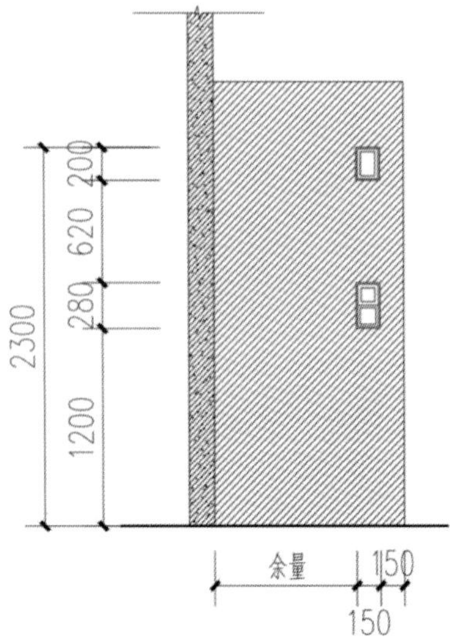

编号 5（门斗外侧）
设备槽尺寸（宽 × 高）：
150mm×280mm
150mm×200mm

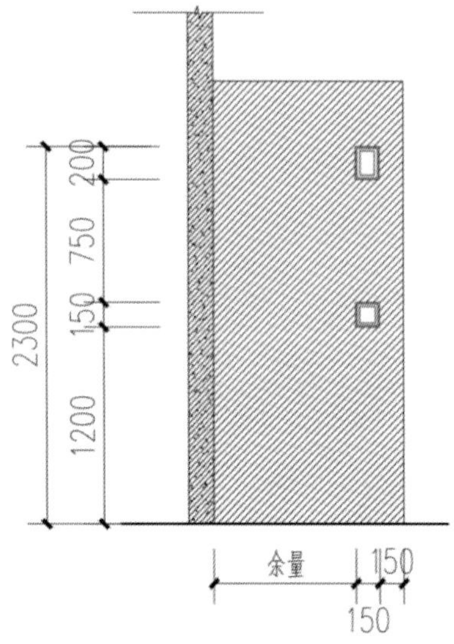

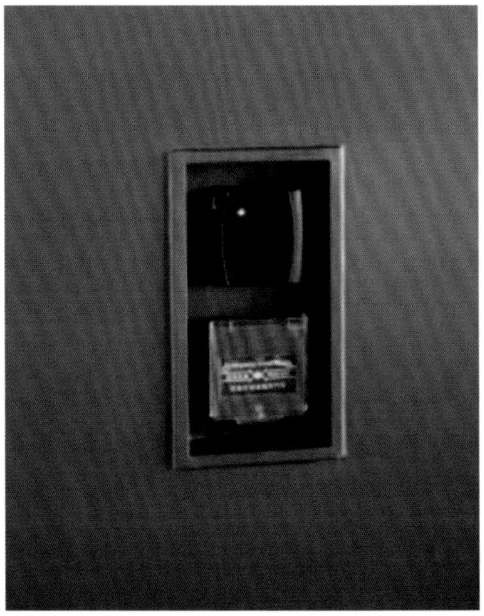

本工程的门口处都设计了金属门斗，门斗上设置了数量及功能各不相同的设备按钮。根据设备专业的设计要求，对门斗上的终端设备进行归类，设计出高度不同的设备槽，设备槽的下部距地高度一致，宽度一致，所有终端设备置于槽中，整洁美观。

5. 全程化服务

本工程结构复杂，驻场设计师面对现场出现的实际问题，需要协调各个专业，及时与业主、施工方等相关团队进行沟通，给出解决方案。

竣工照片

施工过程 1

施工过程 2

竣工照片

施工过程 1

施工过程 2

竣工照片

施工过程 1

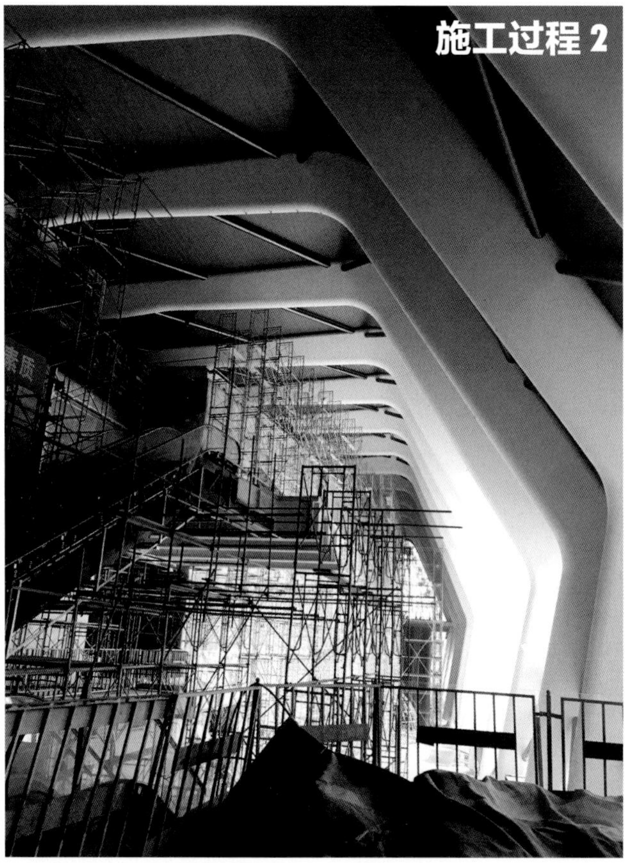
施工过程 2

施工过程 3

施工过程 4

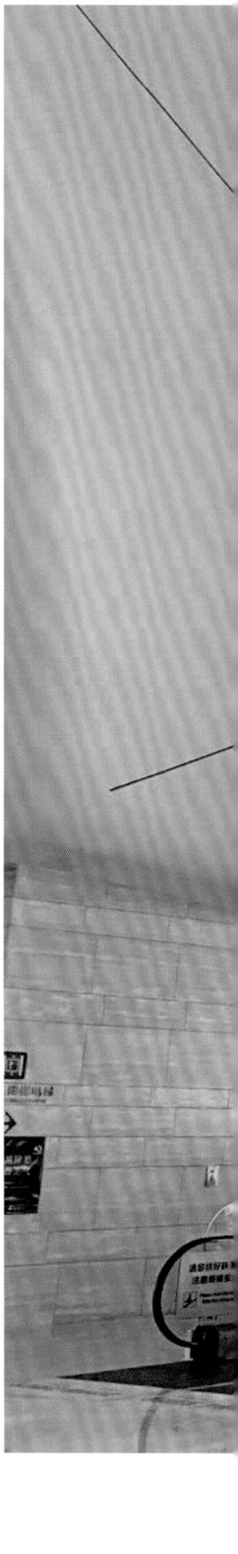

竣工照片

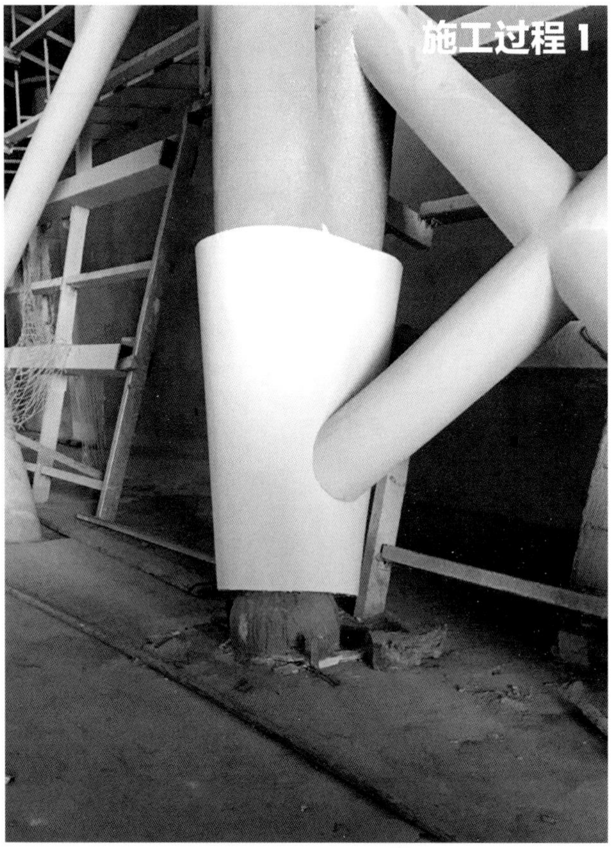
施工过程 1

施工过程 2

施工过程 3

施工过程 4

竣工照片

施工过程 1

施工过程 2

施工过程 3

施工过程 4

竣工照片

施工过程 1

施工过程 2

施工过程 3

施工过程 4

施工过程 5

改善城市环境 共创美丽滨海

竣工照片

施工过程 6

施工过程 7

施工过程 8

二、展览空间

1. 功能分区

根据《博物馆建筑设计规范》（JGJ-66—2015）第 1.0.6 条的规定：“博物馆建筑设计必须与完整的工艺设计相配合，满足藏品的收藏保管、科学研究和陈列展览等基本功能，并应设置配套的观众服务设施。”《博物馆管理办法》（文化部令第 35 号）第十三条规定：“博物馆的建筑设计应当符合国家和行业颁布的有关标准和规范。博物馆建筑应当划分为陈列展览区、藏品库房区、文物保护技术区、公众服务区和办公区等，相对自成系统。” 项目结合国家海洋博物馆的功能定位和世界博物馆的发展趋势及公用设施配套要求，在上述五个基本分区的基础上，增设教育交流区和附属配套区两个分区，形成陈列展览区、教育交流区、公众服务区、藏品库房区、文保技术区、业务研究区、业务办公区、附属配套区八大功能分区。

2. 设计原则

（1）“以人为本，体验为首”建筑与展览空间的设计原则

强调建筑、展陈、展品（文物和标本）与布展艺术及技术应用的有机融合。展厅整体环境色彩沉稳而不沉闷，大气而不琐碎，格调高雅不落俗套。在力求达到主题展览个性化的同时，使其具有专属化的艺术风格 、韵律、节奏和情境，既强调各展览之间内容逻辑内涵的连贯呼应和拓展延伸，又达到多元化展览主题艺术空间与建筑空间环境之间的融合。

（2）建筑空间与展览空间的有机融合

充分利用建筑特有的高低起伏、错落有序的空间布局，统筹展览内容、展品的规划布局，将大型展品、重点故事及其内容做了放大化处理。

同时降低展柜设施、中心展台、辅助展项等基座高度，放大观看视角，营造出不同特性的主题展览个性空间。

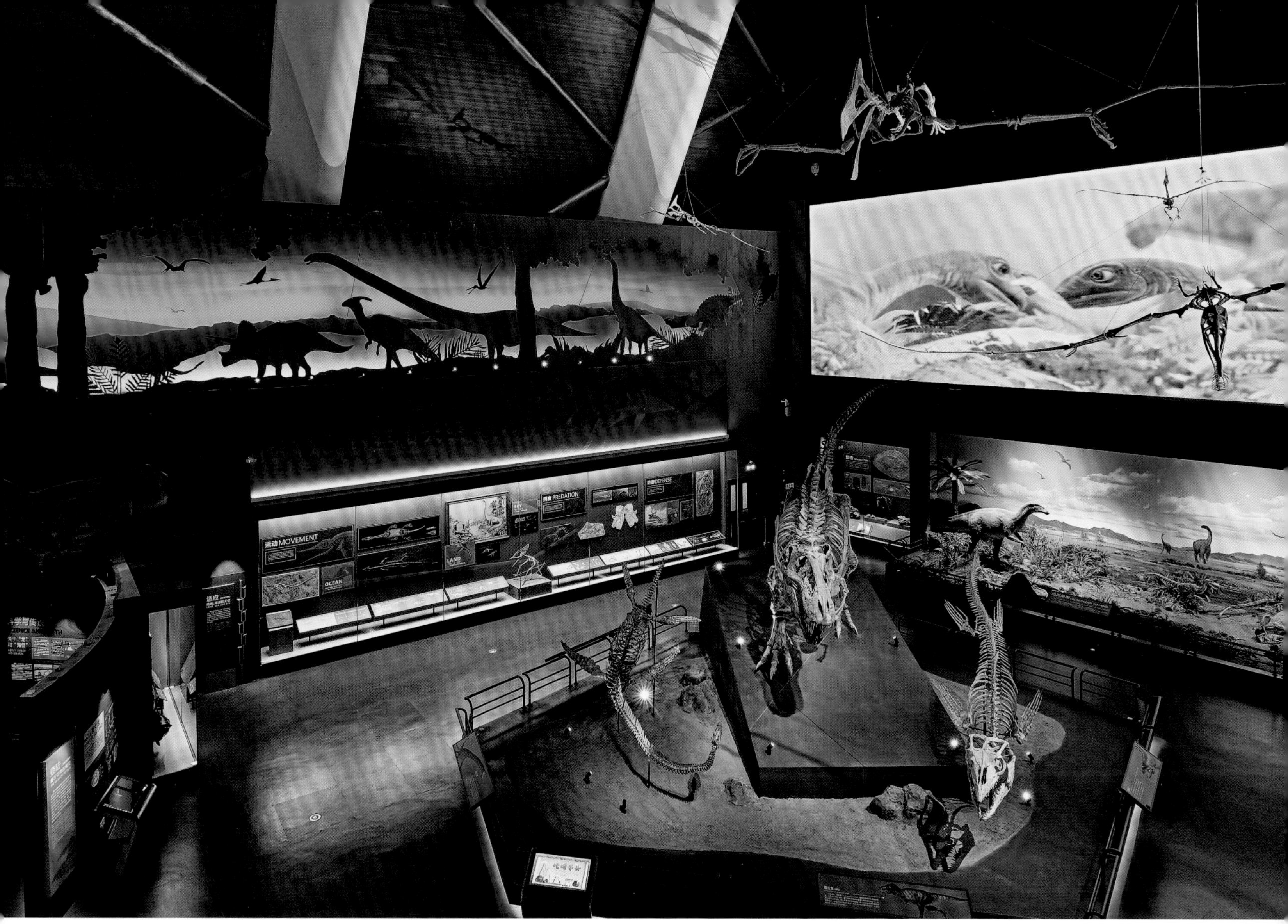

观众服务中心、休息区、餐厅、商店的服务功能及其规划设计风格与建筑空间对接融合，同时将展览元素融入其中，规划出具有个性且提供多种选择的参观路线，以满足自主性参观需求，为观众提供全方位、多元化的观展体验和服务感受。

3. 空间格局

建筑空间格局架构富于变化，全方位满足传统工艺及展览的互动性、体验性、趣味性需求，并为观众提供在有限的时段内选择个人关注和喜爱的内容的可能性，以达到建筑、展览与观众感受的和谐统一。

4. 陈列展览

基本陈列展厅包括：远古海洋、今日海洋、中华海洋文明。

常设专题展厅，包括：龙的时代、航海发现之旅、海洋与天文、筑梦极地、欢乐海洋、从风帆到行轮、蓝色家园、海丝文化主题空间、海洋课堂以及临时展厅。

（1）远古海洋

展厅位于博物馆左翼一层，总面积约 2000 m^2，展厅局部空间挑高达到 11 m，参观展线长约 400 m。该展厅涵盖了从海洋诞生之初至今的各门类海洋生物化石、海洋沉积岩等珍贵展品。

（2）今日海洋

展厅位于博物馆左翼三层，总面积约 2800 m^2，展厅局部空间挑高达 15 m。展览共分为地球海洋、生命海洋与保护海洋三大展区。

（3）中华海洋文明

“中华海洋文明第一篇章”展厅位于 3 号馆二层中部，展览面积共计 985 m^2，层高 12 m，展线长度 180 m。展厅通过实物、展板、模型和多媒体互动等方式，展现在漫长的历史岁月中，中国沿海地区的先民们在认识、开发海洋的过程中所创造出的璀璨中华海洋文明。

“中华海洋文明第二篇章”展厅位于国家海洋博物馆 4 号馆二层南侧，展览面积共计 1195 m^2，展线长度 200 m。展览内容的时间跨度自 1368 年至 1949 年，纵向地介绍了 580 余年内，中国人民海洋观的变化和发展。总的来说，海洋与人、与国家的关系，是共生的。

“中华海洋文明第三篇章”展厅位于 1 号馆二层东侧，由序、认知海洋、人海和谐、依海富国、以海强国、合作共赢六部分组成。展厅面积 950 m^2，展厅层高 3.6 m，展线长度 200 m。展览以全方位的角度，介绍从中华人民共和国成立到 21 世纪的今天，中国在海洋经济、海洋生态、海洋探索等多领域取得的一系列成就。

海洋强军建设
萨镇冰
刘冠雄
清末至民国的海权维护
「收复东沙群岛」
「巡视西沙群岛」
张人骏

探索的脚步从未停息
2013.11.8上述
海龙号
蛟龙号
潜龙一号、潜龙二号、潜龙三号
QIANLONG 1, QIANLONG 2, QIANLONG 3
蛟龙
潜龙
三号

（4）海洋课堂

海洋实验课堂
MARINE EXPERIMENT CLASS
国家海洋博物馆
NATIONAL MARITIME MUSEUM

（5）龙的时代

展厅位于 2 号馆二层南侧，展厅以中生代爬行动物为主题，以海洋、陆地、天空三个维度为出发点，展示了距今 2.52 亿 ~0.66 亿年期间爬行动物的演化历史。展厅面积 1000 m^2，建筑层高 3.6 ~ 15 m，展线长约 200 m。

（6）海洋发现之旅

展厅位于2号馆三层南侧，包括：序厅、达尔文的环球考察、南方大陆、极地世界、非洲之旅、深入新大陆六大板块。展厅面积 2100 m^2，建筑层高 9 m~12 m，展线长约 350 m。

（7）海洋与天文

展厅位于博物馆三层东侧，展厅局部空间挑高达 10 m，面积 1620 m^2。该展厅是目前国内首座将海洋、航海、天文、航天进行有机融合的主题展厅。

（8）筑梦极地

展厅位于4号馆三层北侧，展览以“生命与环境、探索与发现、极地与人类”为主题，分为三个展区：极地生物、科考基地、极地生存。展厅面积860 m²，建筑层高9 m，展线95 m。

（9）欢乐海洋

展厅位于 3 号、4 号馆三层东侧，主题空间布展面积约 1500 m^2，整个空间分为热带海洋、温带海洋、极地海洋三个环境分区，以及相应的过渡区域。

（10）从风帆到行轮

展厅位于 4 号馆一层东侧，展厅空间高度 3.5~9 m，展览面积共计 780 m^2，展线长度 98 m。运用高低错层展览空间及展示手段，感知近现代船舶历史的基本轮廓和发展路径。

（11）蓝色家园

展厅位于 4 号馆一层北侧，展厅面积 1280 m^2，展线 95 m。海洋孕育了地球最初的生命，是无数海洋生物的家园。

（12）海丝文化主题空间

展厅位于 3 号馆一层中央大厅右侧，呈环状展线布局，展厅面积 2549 m^2。其中福船总长 27.8 m、高 22.9 m、宽 8.6 m，位于建筑空间最大高度 28 m 光环境互动的展位。参观者可直观地了解我国古代辉煌的造船、航海技术以及对外贸易情况。该展示成为整个展区的核心。

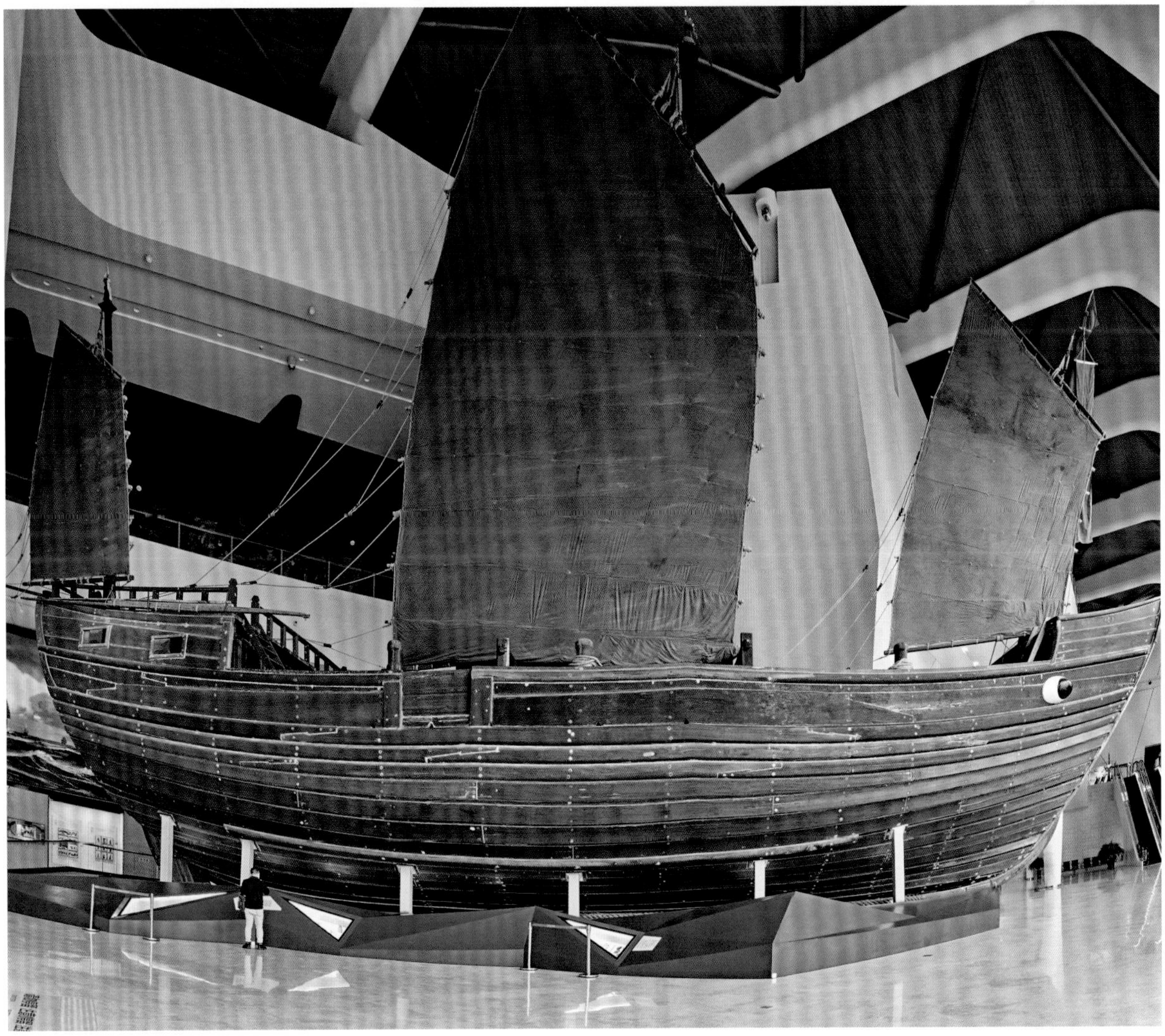

三、结构设计

国家海洋博物馆的建筑造型独特、美观，简洁流畅柔美的曲线语言令参观者产生对海洋元素无限的遐想。这样创新的独特形态，需要设计与结构专业的无缝对接才能完美实现。该建筑由中央大厅和四个展厅组成，各厅长宽高分别约为：1 号厅，长 190 m、宽 33 m、高 25 m；2 号厅，长 248 m、宽 34 m、高 30 m；3 号厅，长 245 m、宽 40 m、高 33 m；4 号厅，长 180 m、宽 30 m、高 25 m。1、2 号厅以及 3、4 号厅分别通过两个不同宽度的共享公共走廊连通。2、3 号厅之间是约 21 m 高的入口中央大厅。4 个厅的楼面标高分别为 6 m、12 m、18 m、22 m，展览区域柱网为 12 m × 12 m。依照建筑立面造型的创新性要求和满足展陈布展设计弹性空间的需求，结构设计采用大跨度无柱空间设计、超高跨层空间设计、超长悬挑空间设计等结构体系的空间整合，并且全过程运用 BIM 技术，实现结构力学与建筑美学的高度统一和完美融和。

1. 主体设计

引发人们无限遐想的 4 个主展厅，长度为 180~248 m，高度为 25~33 m，跨度为 30~40 m，均为超长结构。其屋盖和外墙为不规则整体曲面，并有机连成一体，每个厅的内部在不同位置均设有通高的展厅，这就让曲面屋盖和外墙既是建筑围护结构又是内部的装饰部分。

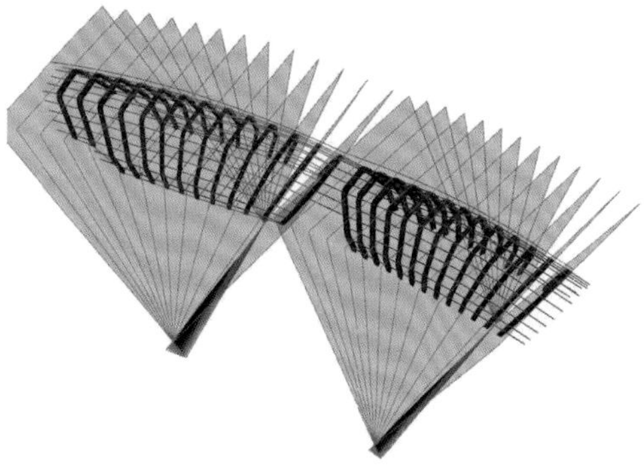

（1）不规则的弧形屋盖结构

为更好地实现四个主展厅的屋盖和檐墙连体建筑外形的非线性双曲面形态，结构设计在水平与垂直方向按一定角度设置具备良好结构特性的 112 榀门式平面钢桁架，连续布置成为一个有机的整体结构空间。门式钢桁架的上弦间均设置管状钢斜撑，构成网状的整体结构，为大厅长向提供抗侧力稳定性，同时减小门式钢桁架的屈曲长度。钢桁架根部间距为 9~15 m。

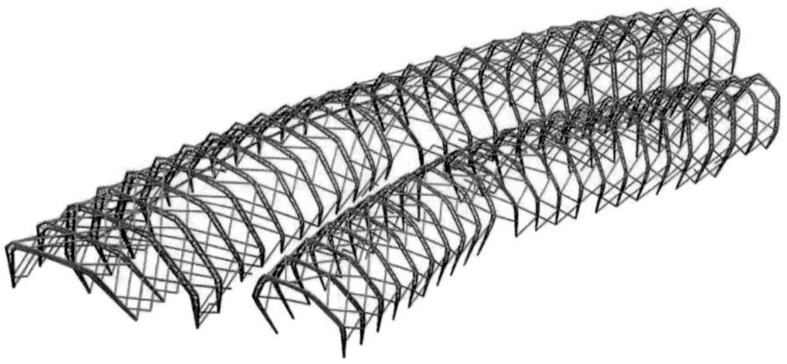

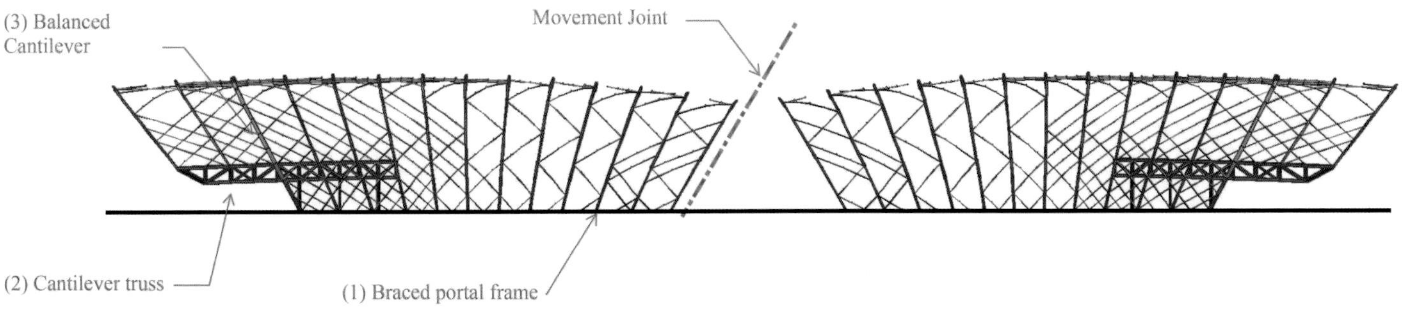

门式钢桁架结构示意（组图）

（2）门式钢桁架设计

门式钢桁架采用具备良好结构特性的平面桁架结构，每榀平面桁架由空心圆钢管相贯构成。门式平面钢桁架的倾斜布置，使门式钢桁架间的斜杆在正常工况下均受拉力，这就减少了其截面尺寸，更好地实现了建筑效果，并节省材料。每榀门式平面钢桁架由钢管相贯组成并外包 GRG 装饰板，建筑装修设计与结构的完美结合，使结构构件成为内部装饰的一个元素。有人说它像古船的内舱骨架，让展厅内的游客有置身于船舱中的感觉。

（3）中央主入口大厅屋顶层结构布置

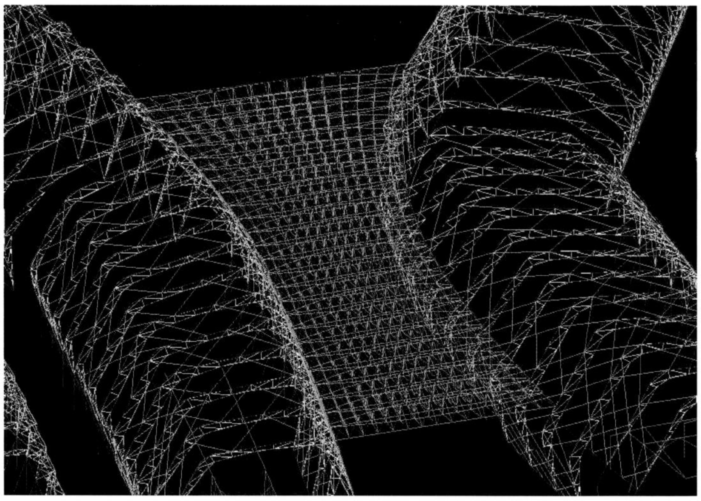

中央大厅屋盖由透镜形状空腹桁架组成，支撑在 2 号和 3 号大厅的门式钢桁架上。空腹桁架由钢管拼成梭形桁架由钢斜撑提供平面外稳定性。支承体系架构呈锯齿形，支承设计为一端铰接，一端为限位滑动的支座，即容许屋盖结构有一端是可以有水平位移的。这样设计使屋盖在温度、风和地震作用下，大厅的支撑门架仅产生竖向及水平力，使主体结构受力简单，结构更加安全。

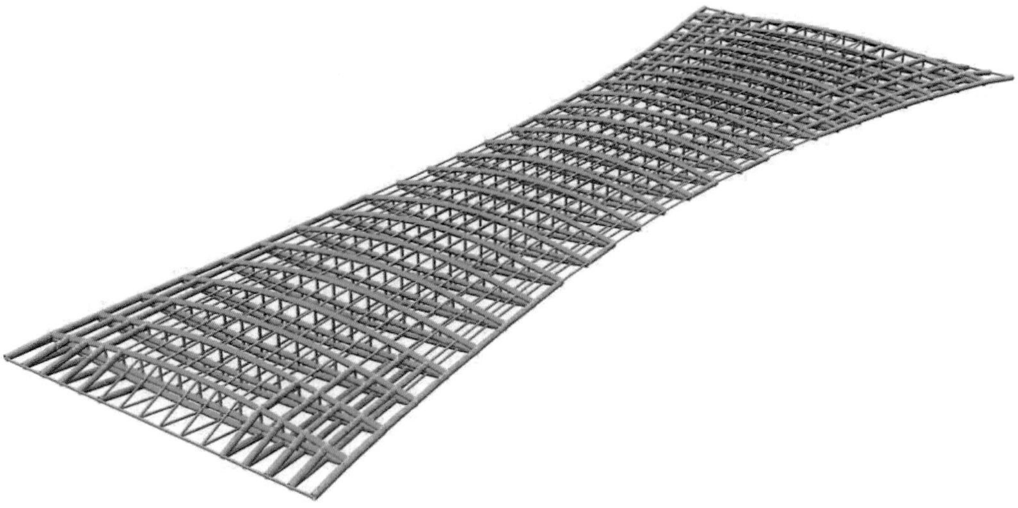

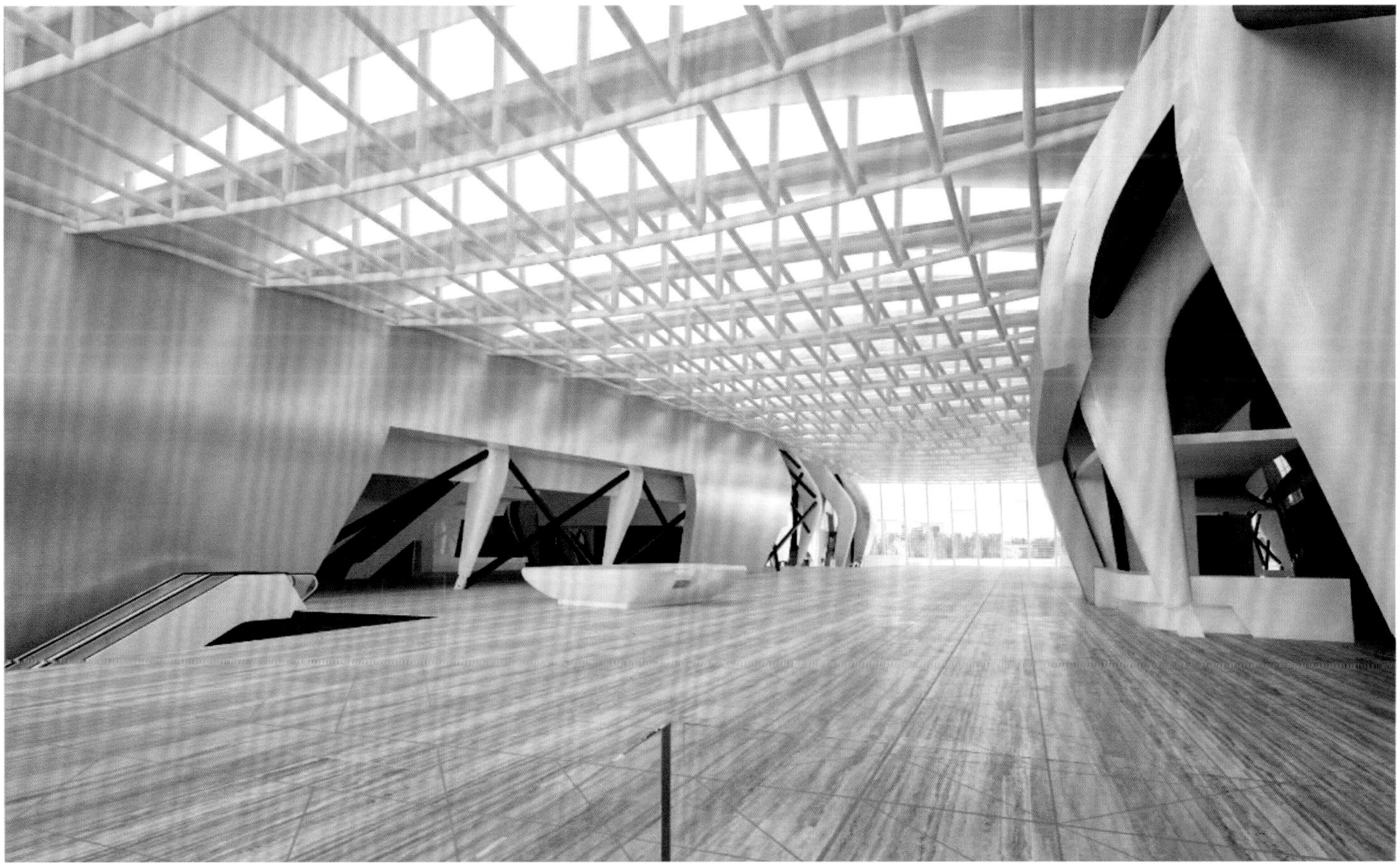

（4）支座设计

由于每榀桁架在水平与垂直方向均不垂直，为解决结构受力、变形及安装问题，在桁架脚部设置能转动的万向抗震球铰支座。球铰支座是结构关键部位的一个重要构件，其最大可承受 16000 kN 压力和 7800 kN 剪力的共同作用，还可保持 0.1 rad 的转动。

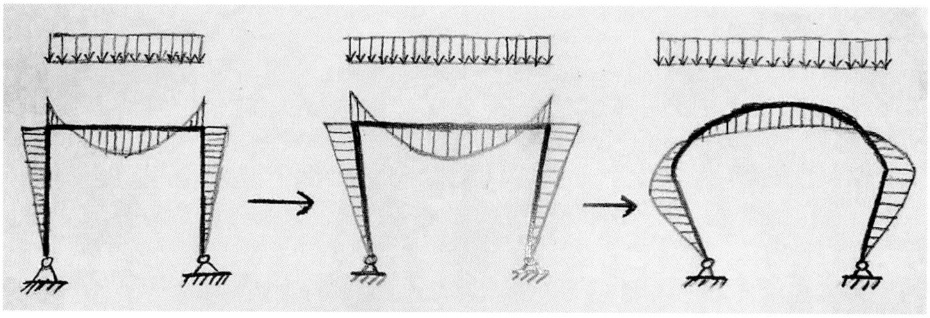

对其用 ANSYS 软件进行有限元分析，从结果看，加载 1.0~1.5 倍设计荷载时，支座荷载位移曲线呈线性，竖向压缩位移、应力等均能满足《桥梁球型支座规范》（GB/T17955—2009）的规定。

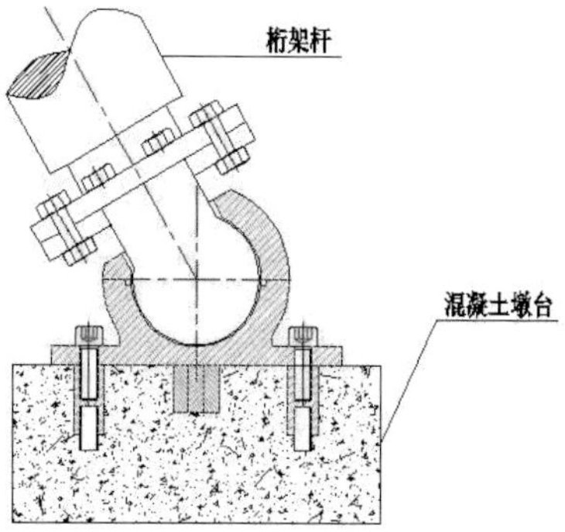

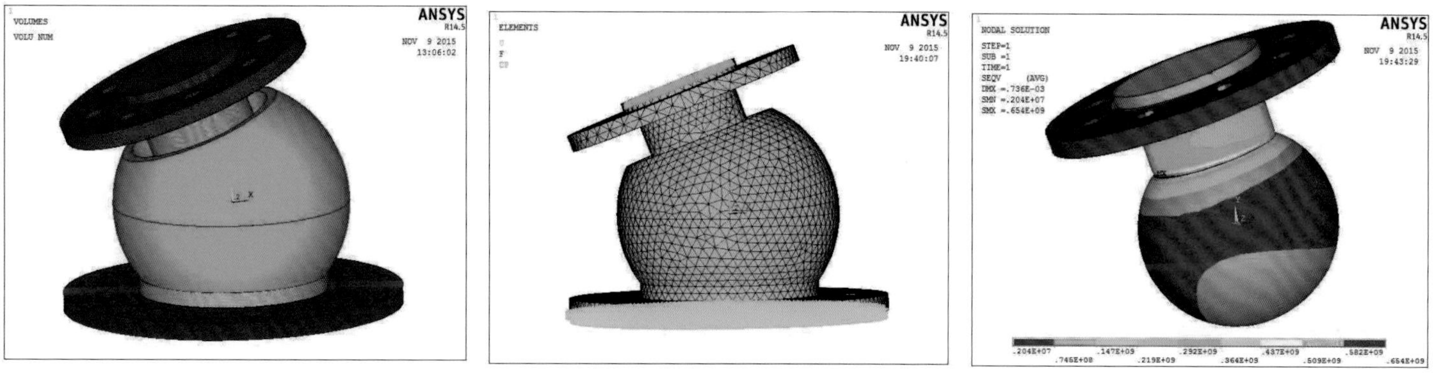

对关键节点、变截面铸钢件等，用 ANSYS 软件进行多工况受力分析，以保证关键结构构件的安全度，这种合理连接设计保证了杆件间内力传递顺畅。

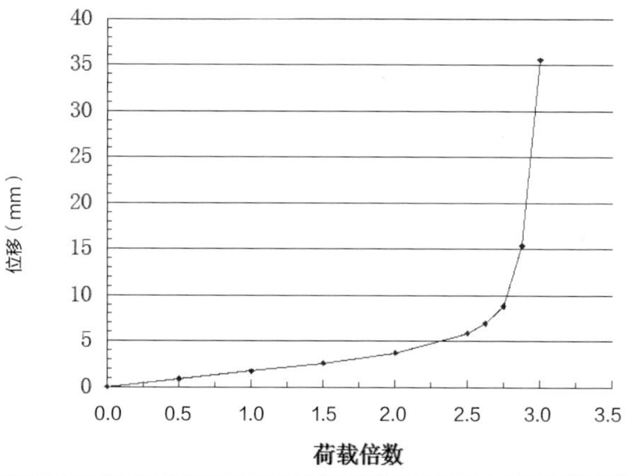

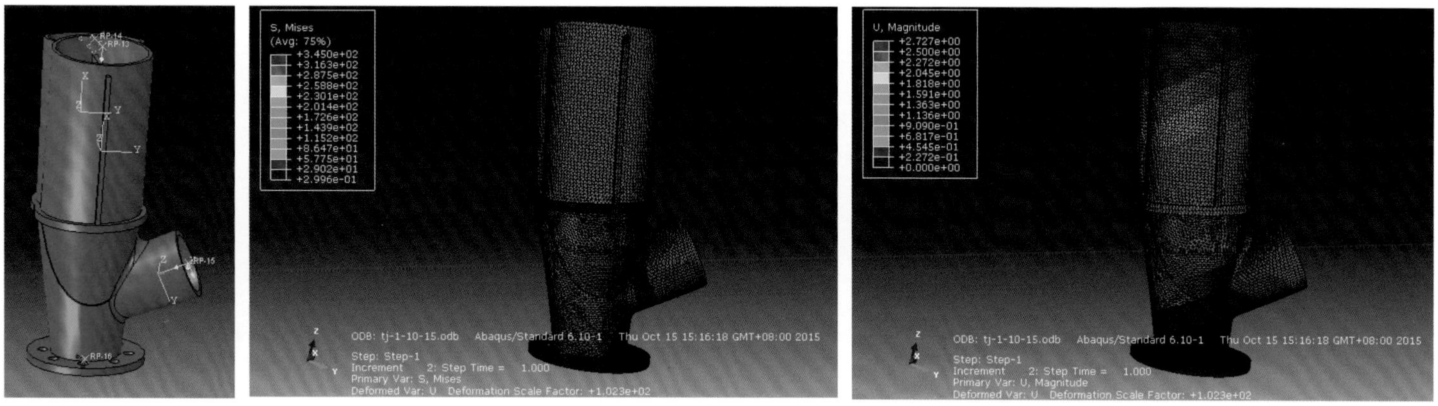

节点结构分析（组图）

（5）支座的检测

选取了1 ：1的实体球铰支座，使用压力机进行加载试验。通过试验得到的荷载—位移曲线，结果显示曲线基本呈线性，满足设计与规范的要求 。此种新型抗震支座的设计，获得了国家实用新型专利。

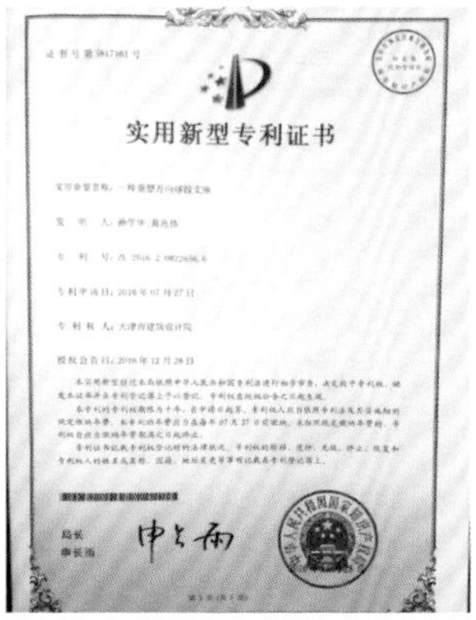

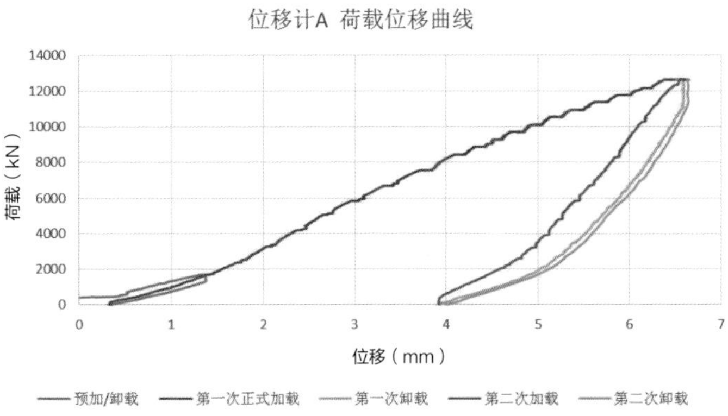

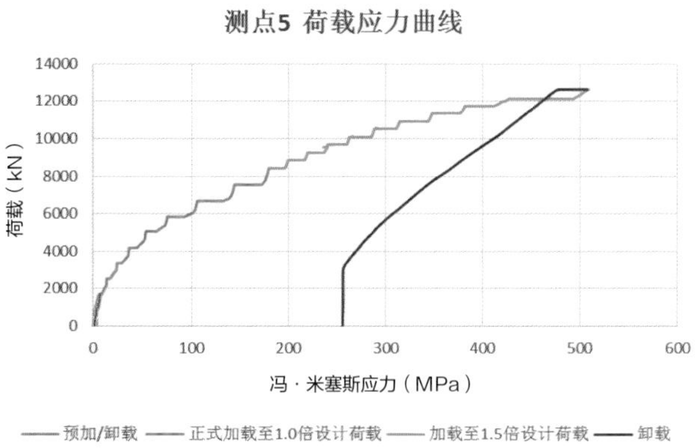

（6）悬挑结构设计

设在在海陆交接处的国家海洋博物馆，其主馆大厅的南北尽端是悬挑于水路海面和陆路广场之上的，这让参观者有一种站在船头航行于海面上的感受。

对此大尺度悬挑结构，设计以底部大型悬臂钢桁架和上面、侧面门式钢桁架间的斜撑支撑组成一个整体的结构空间体系，既保证悬挑部分的整体稳定性、安全性，又彰显悬挑建筑轻巧灵动并与大厅空间浑然一体。

巨大的楼面结构悬挑长度达 39 m，楼板厚 150 mm。在悬挑结构的外侧，曲线形外围护屋面和墙身又向外悬挑 8~18 m，屋面最大悬挑长度达 55 m，形成了具有视觉冲击力的沿海建筑景观。

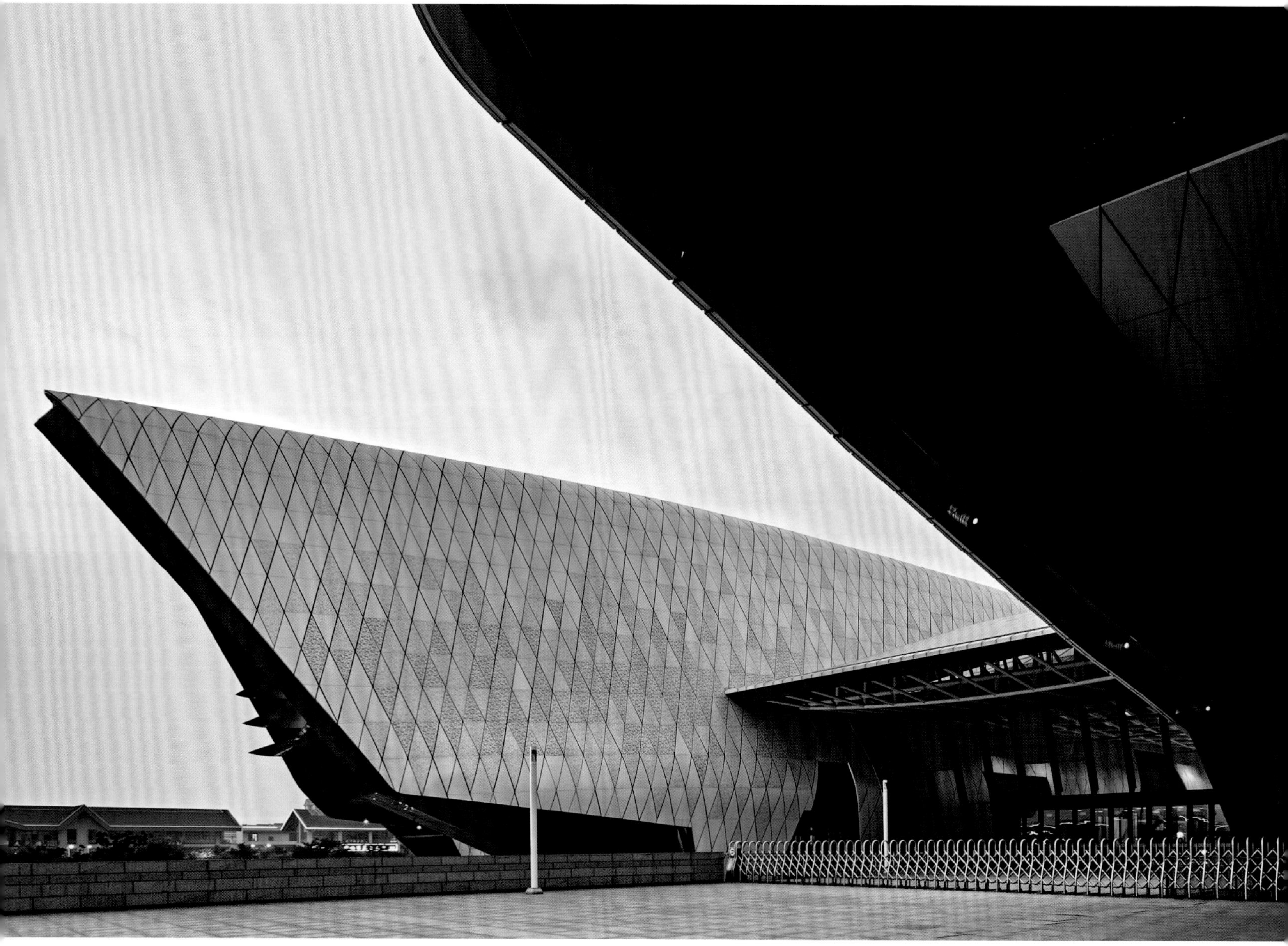

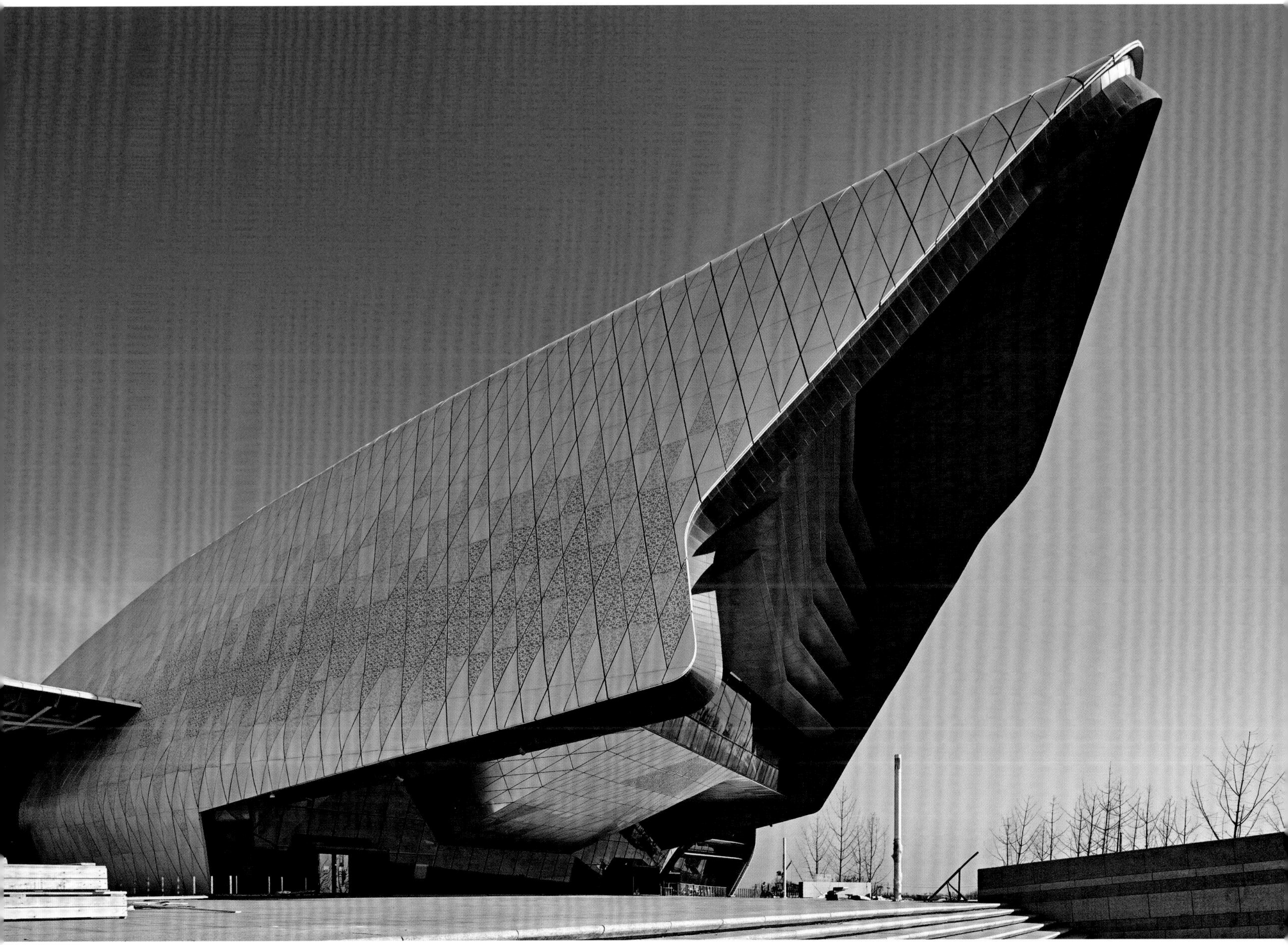

2. 复杂的地基设计

（1）场地条件

本工程坐落在渤海之滨的中新天津生态城南湾的海域吹填区，场地原为浅海，建设前期人工吹填成陆地，地表存在6m左右的填土层。博物馆为临海建筑，地基土和地下水均有腐蚀性，这给博物馆的基础设计带来较大技术难度。

（2）基础设计

为保证博物馆内藏品的安全，建筑设计有效地将建筑整体抬高，建筑的首层地面高于场地地坪，首层地面至场地地坪之间设一架空层，将架空层作为设备管道层，既避免了地下水对建筑的腐蚀性侵害，也解决了填海地面的沉降给建筑带来的隐患。

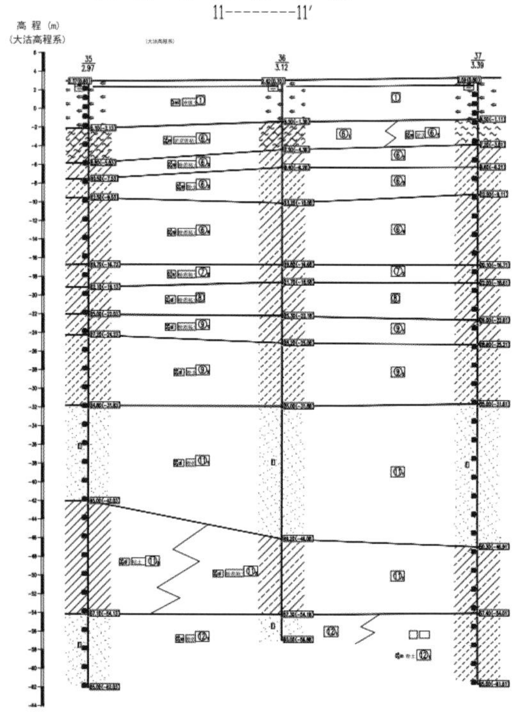

工程地质剖面图
11————11′

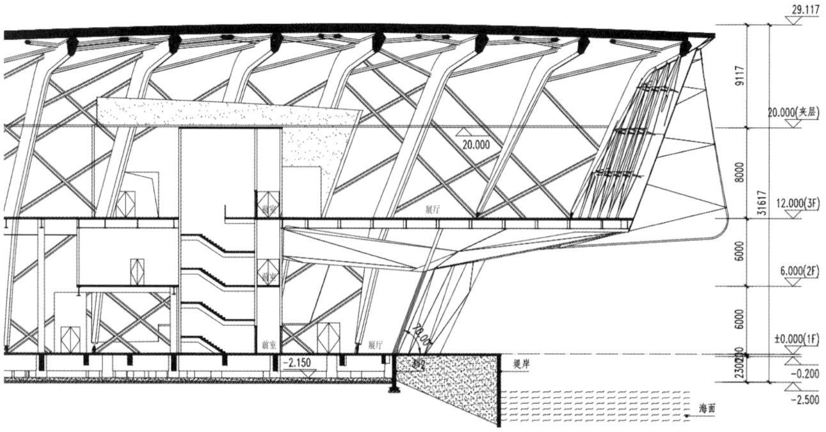

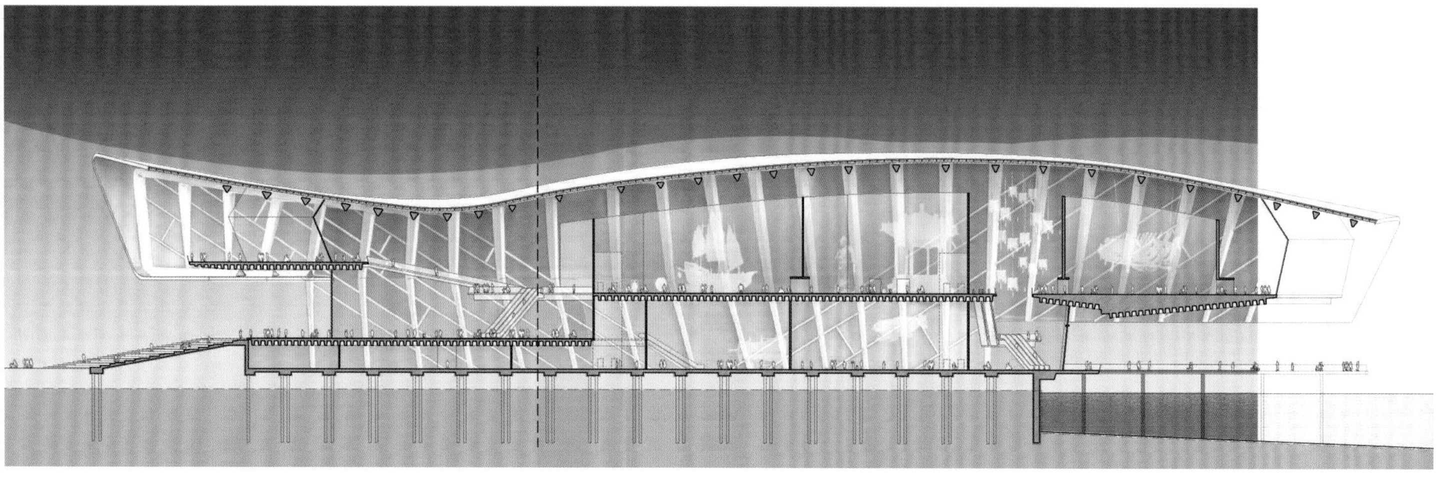

结构系统简图

3. 结构全过程三维数字设计

（1）全参数化建模

考虑非线性双曲面造型的空间、内部无柱大空间，超长的悬挑结构等需求，若采用传统的结构体系及设计方法来展示灵动的建筑内外空间，是根本无法实现的。为了完美地体现设计创意的效果，结构选型、结构布置、三维分析、施工图设计、施工实施全程均运用三维数字化技术，实现了从方案设计到工程竣工，再到后期运维管理全过程三维数字化技术全覆盖，充分体现了博物馆设计的高完成度和科技含量。

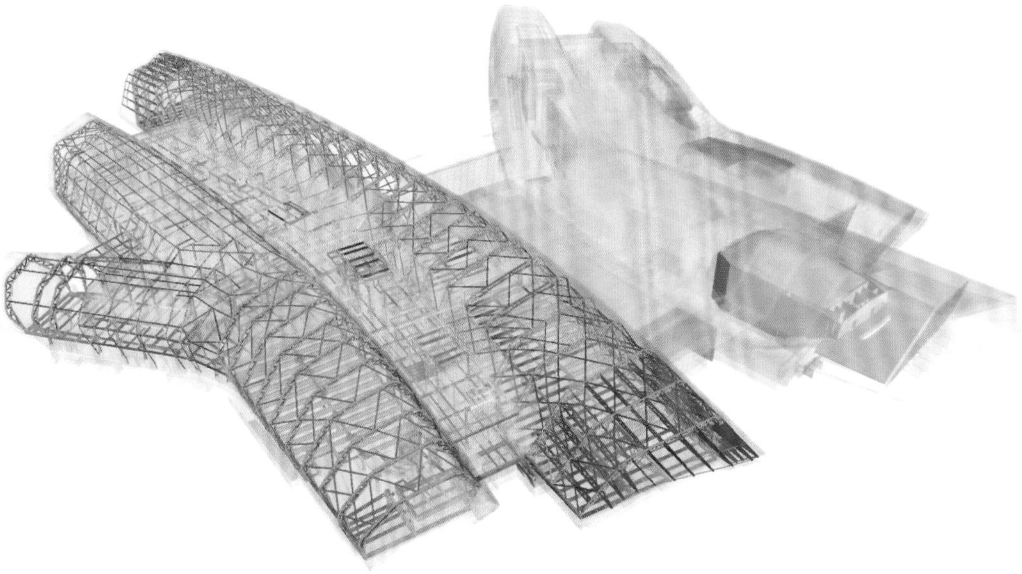

（2）实体三维模型试验

在方案阶段，利用 RHINO 软件实现在复杂建筑外表面上建立合理结构模型的结构定型工作，并通过参数化优化，反复调整建筑与结构模型，以实现最合理的造型对接，这是传统的二维设计无法实现的。

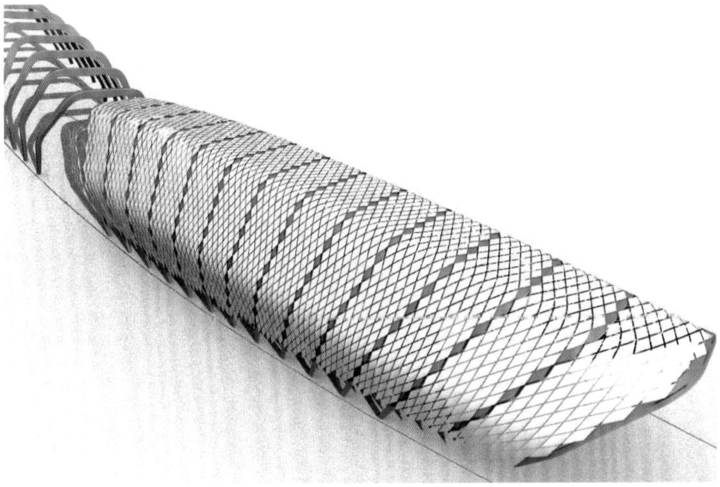

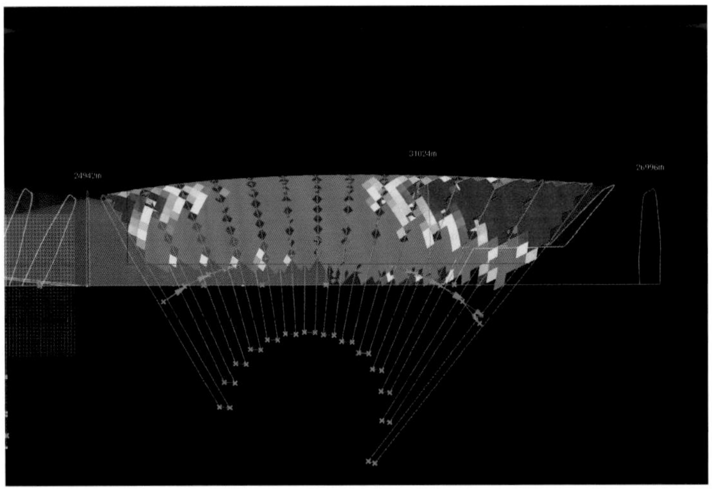

（3）严谨的计算书分析

在确定外形时，结合建筑平面建立内部主体结构的三维数字模型，为后续结构的地震、积雪荷载、风荷载分析确定基准模型。

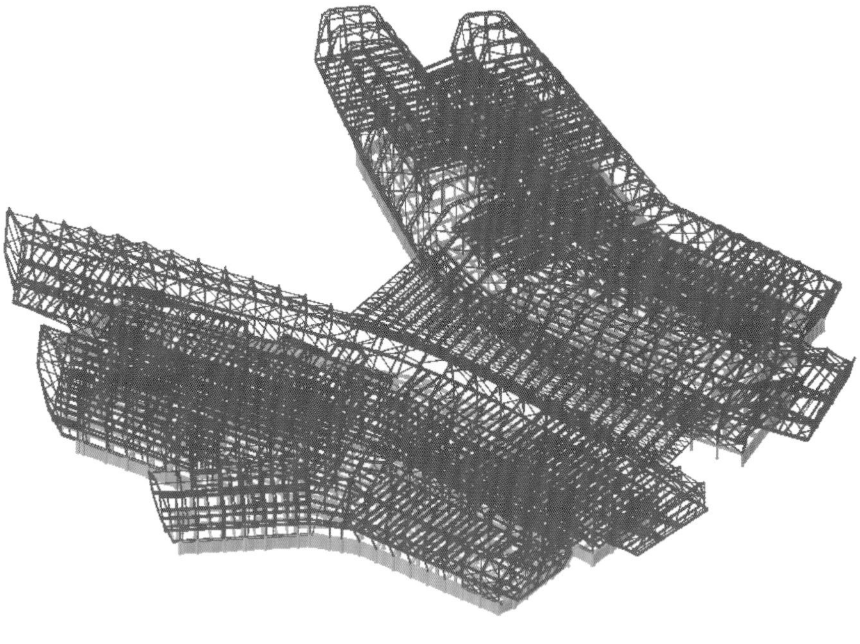

按主体结构三维数字模型制作 1 ： 200 实体模型，送风洞试验室进行全天候风环境模拟试验，以科学确定合理风压、风振等计算参数。

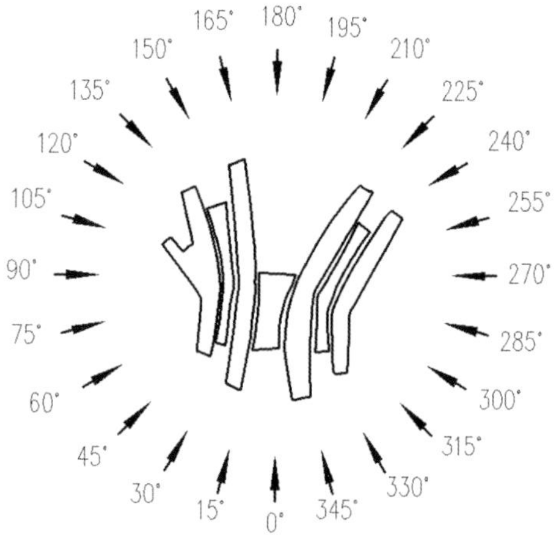

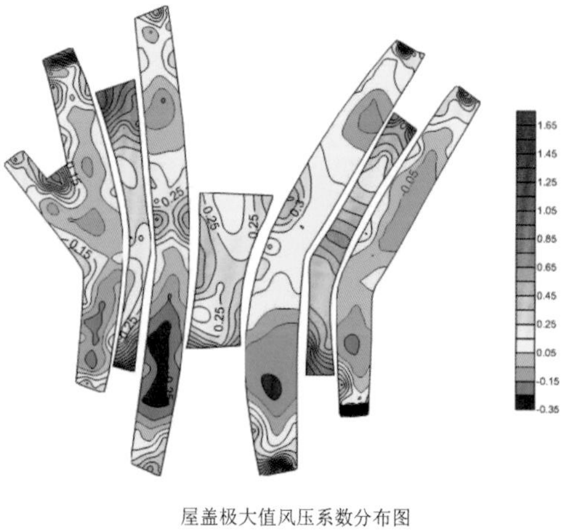

屋盖极大值风压系数分布图

（4）全数字化制作安装

分别采用国际通用有限元结构分析软件 Midas/GEN、国内的 YJK 等软件，对正常使用时，结构在恒荷载、活载、全向风载、不对称雪荷载（半跨）、温度荷载等作用下的响应进行验算，并利用反应谱法进行多遇地震分析，以确定最优的结构方案，达到安全、经济的目标，实现最好的建筑效果。

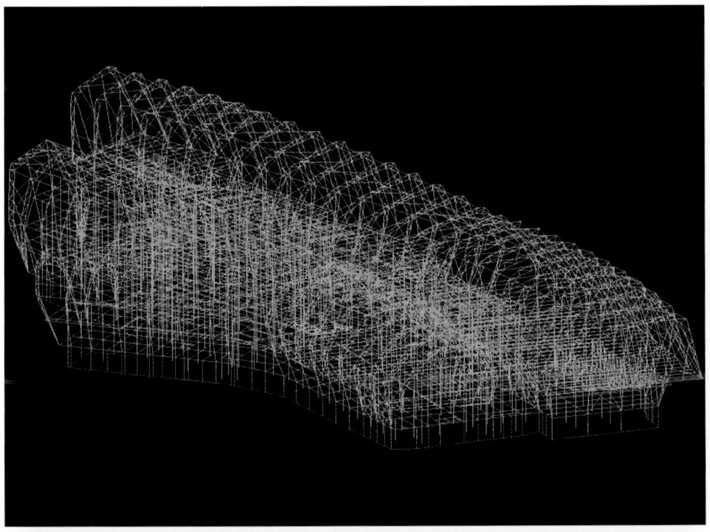

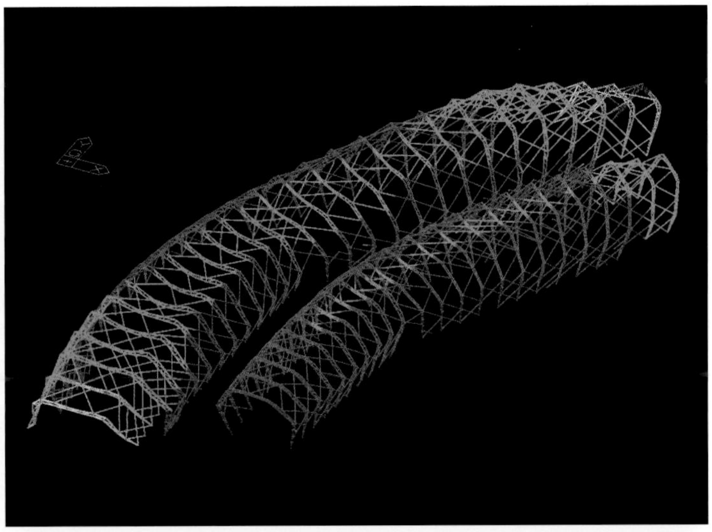

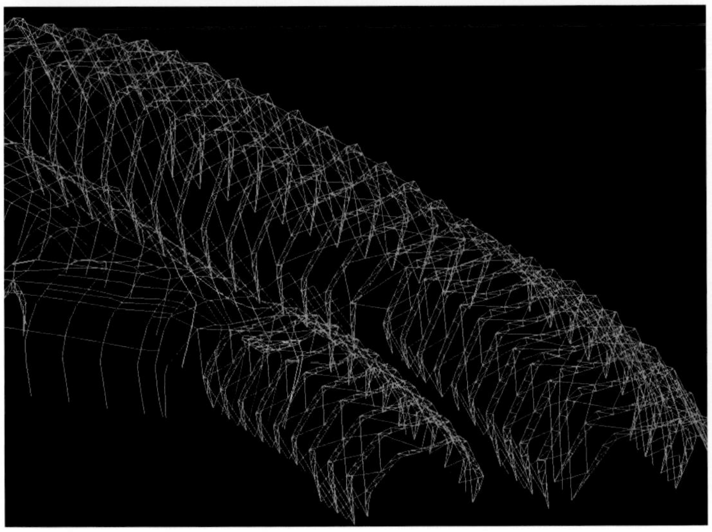

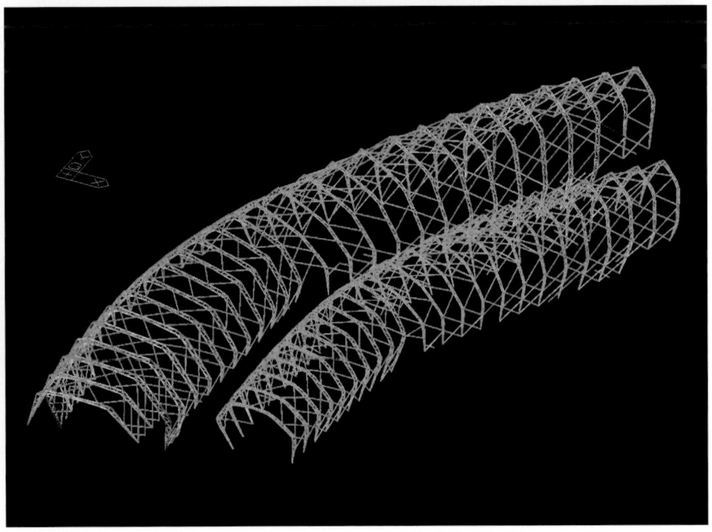

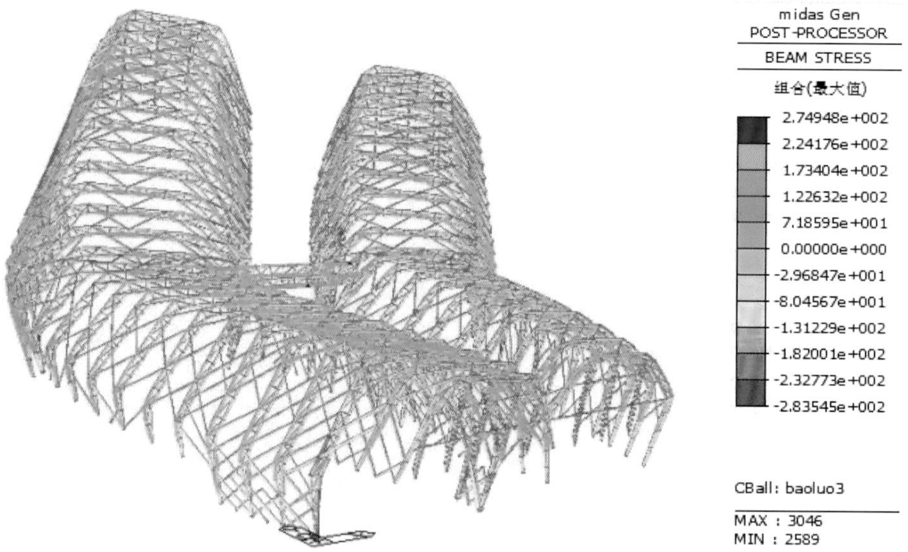

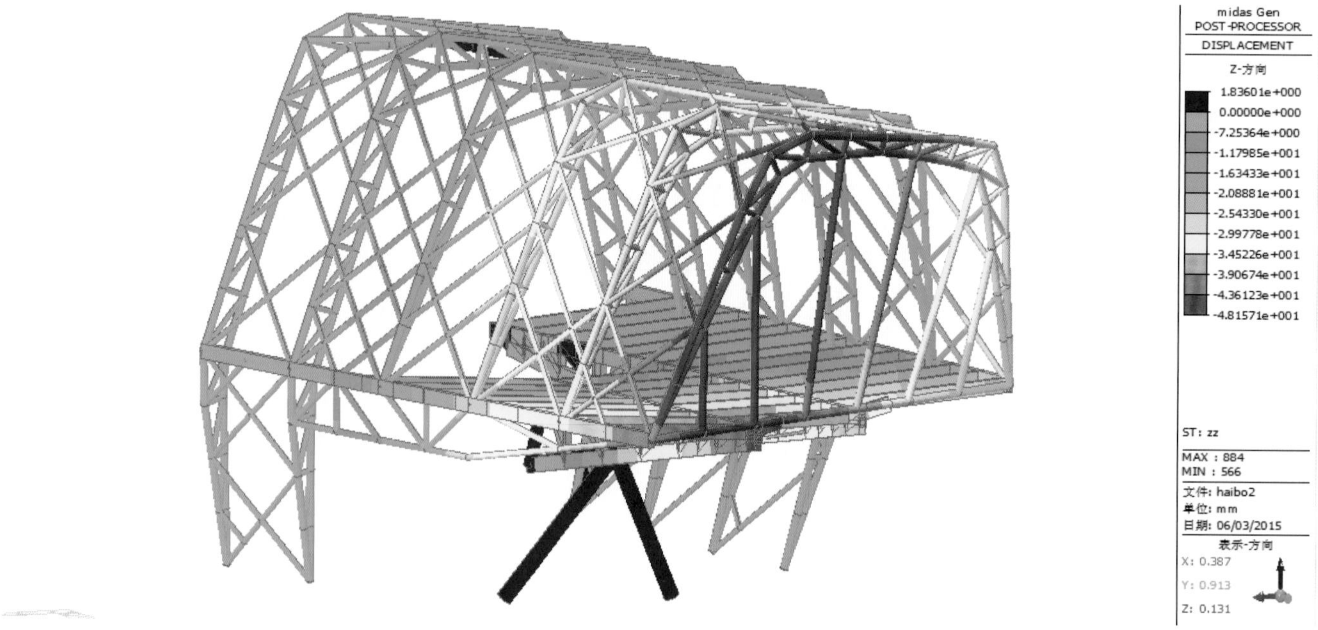

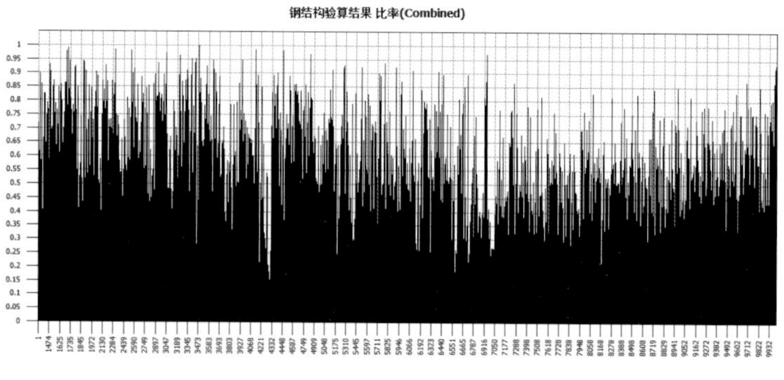

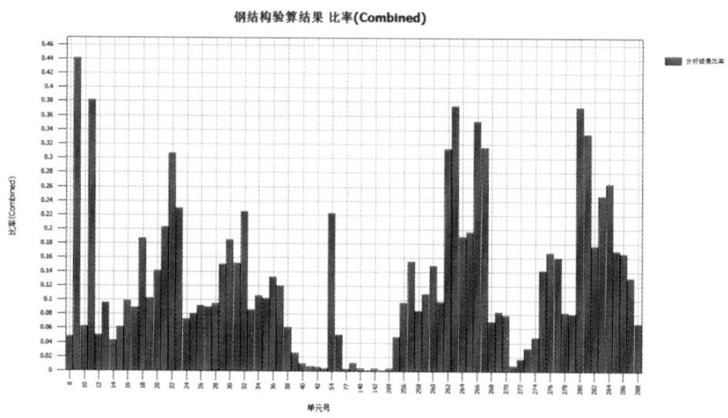

利用 Midas、SAP2000 等三维空间分析软件对主体结构进行多模态全工况下的结构分析计算，确定安全合理的结构布置、安装施工顺序。

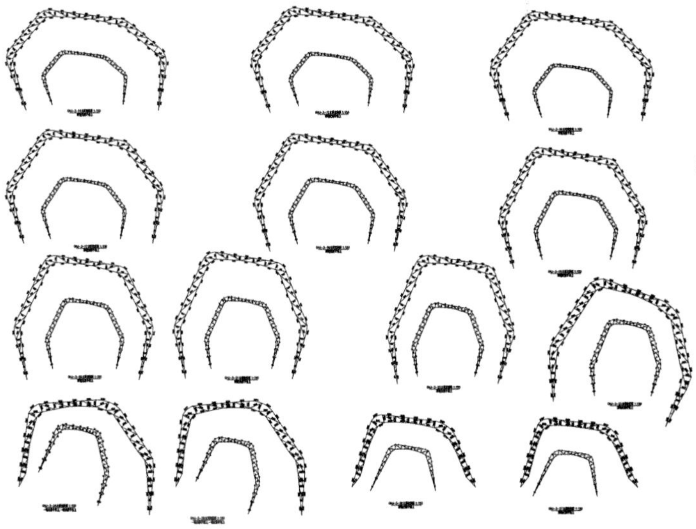

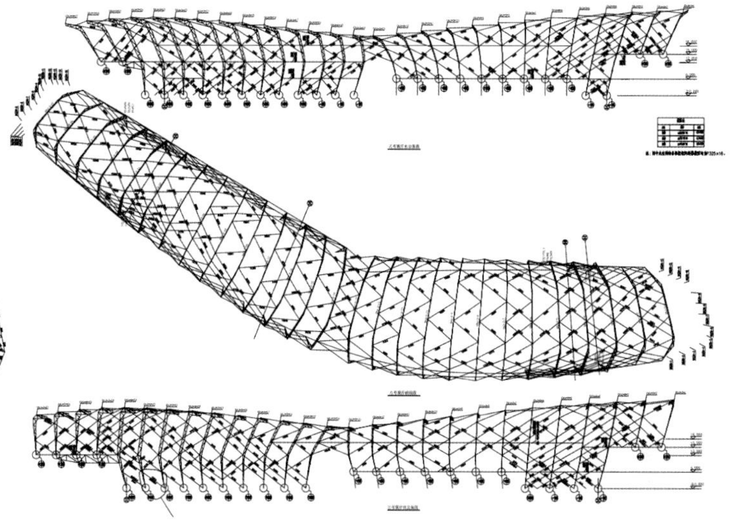

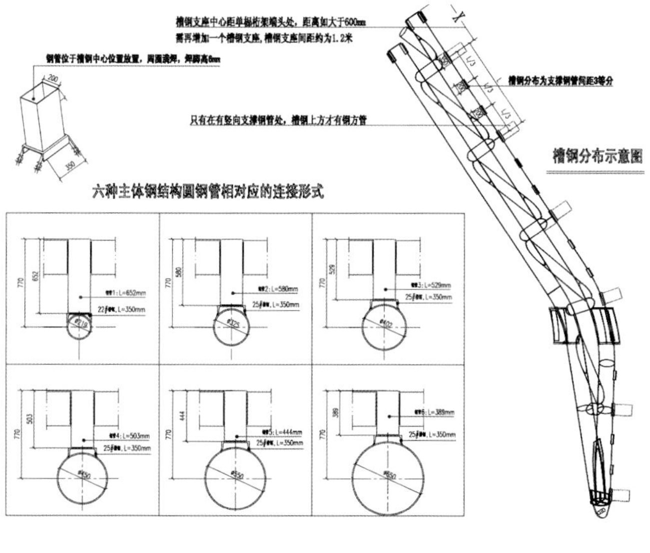

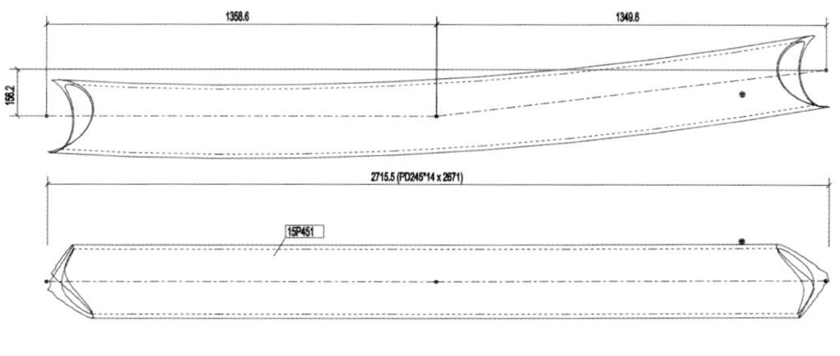

e. 全数字化监测检测

利用 TEKLA 等后期处理程序，对主体结构进行拆分，形成全数字化三维全息施工图，并与二维纸质施工图交互使用，以指导施工、验收、结算等工作。

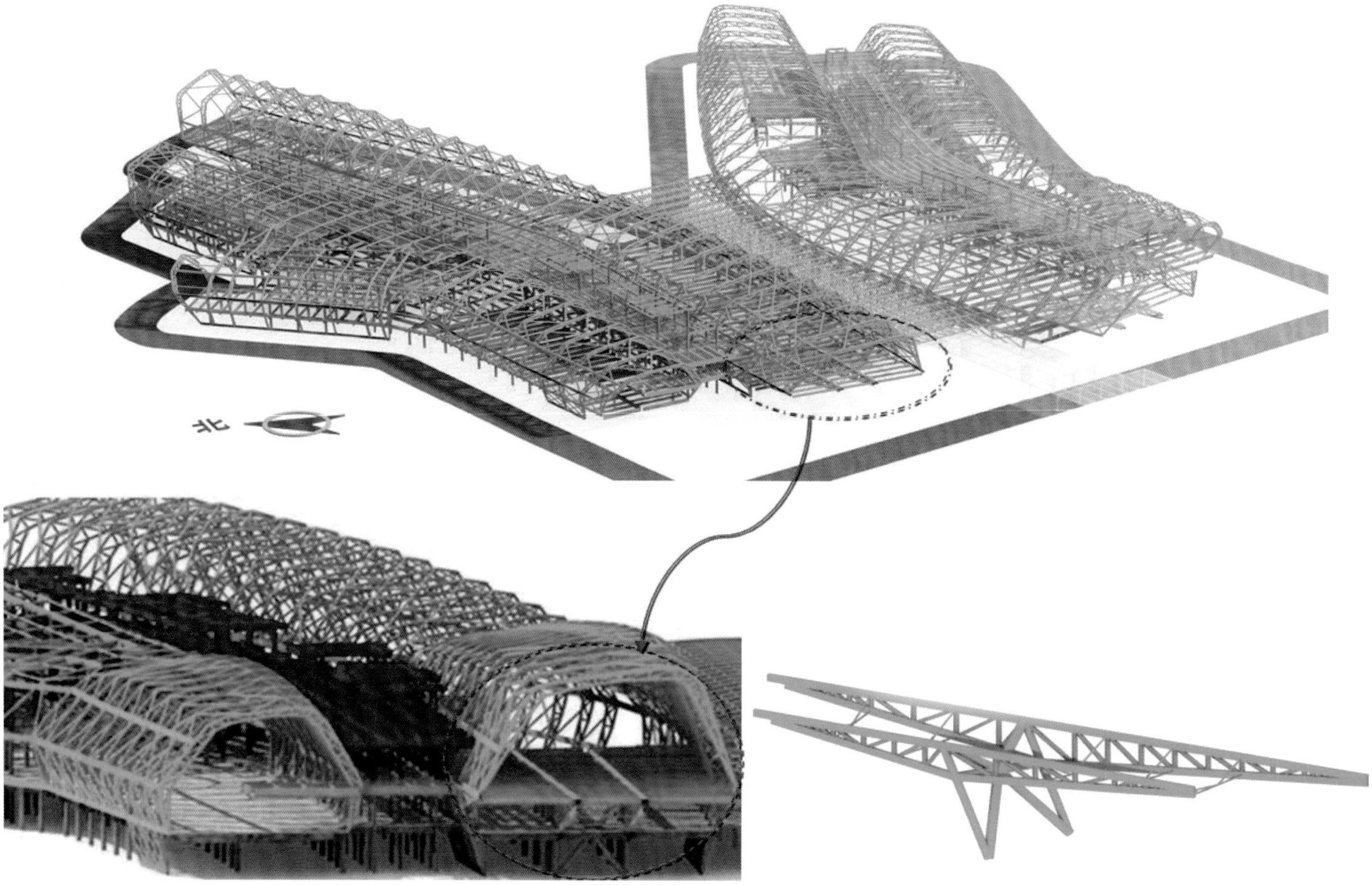

楼层钢桁架效果图

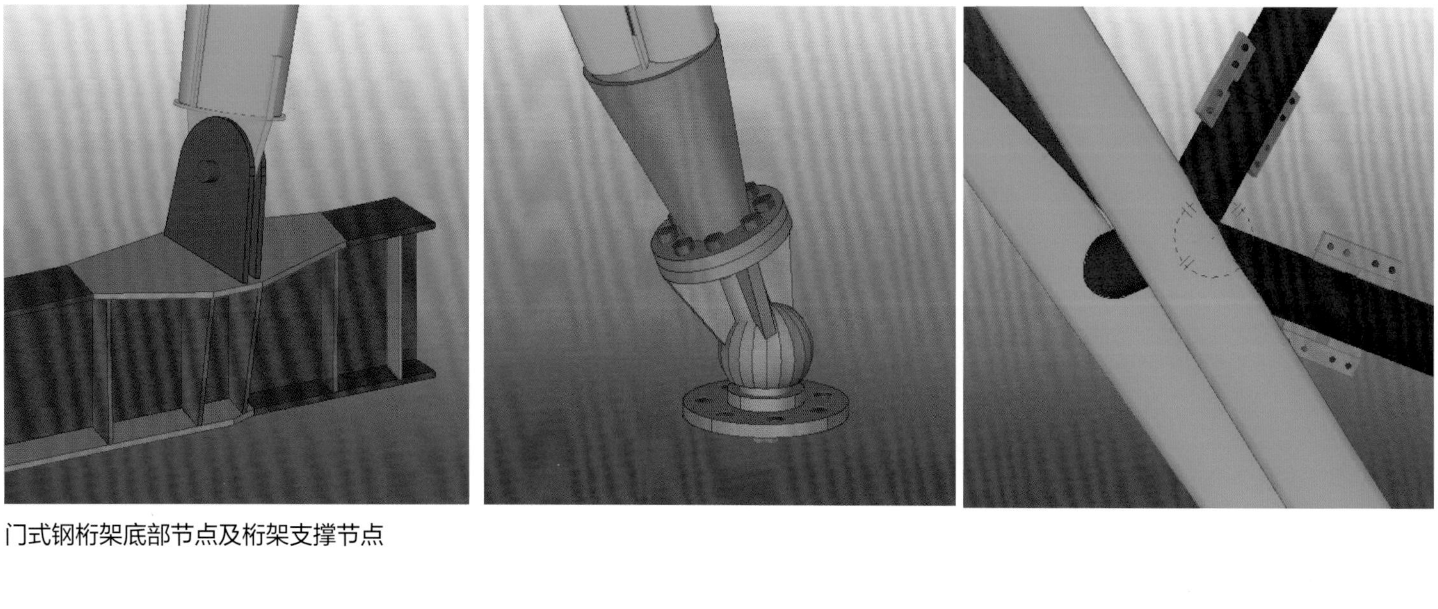

门式钢桁架底部节点及桁架支撑节点

在施工过程中，全面利用三维数字化模型，并用全站仪等设备对主体结构进行全过程检测、监测。

施工模拟

施工模拟

带有二维码标识的钢结构构件

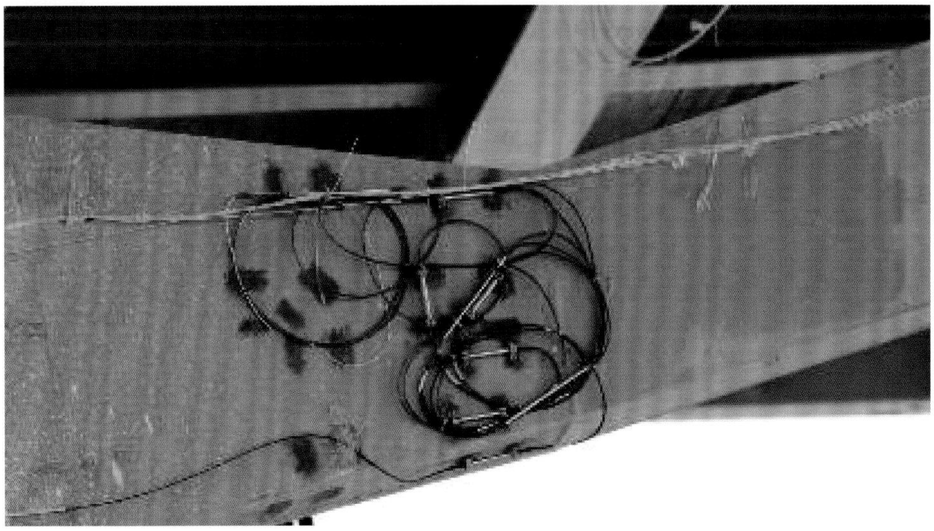

电子监测

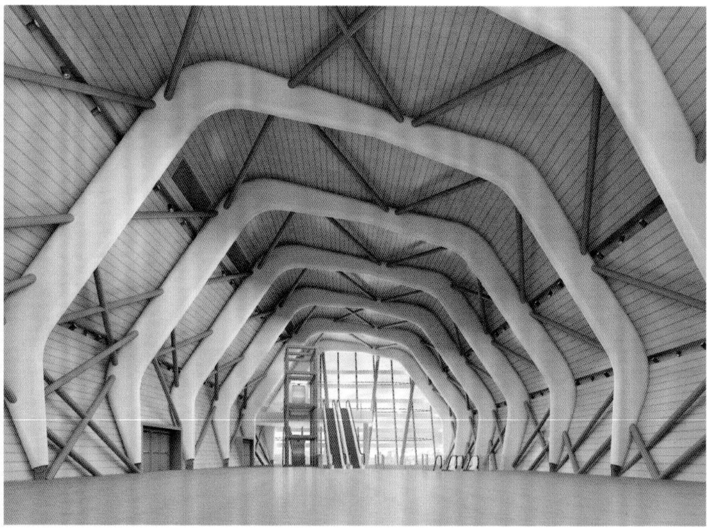

主体建造过程（组图）

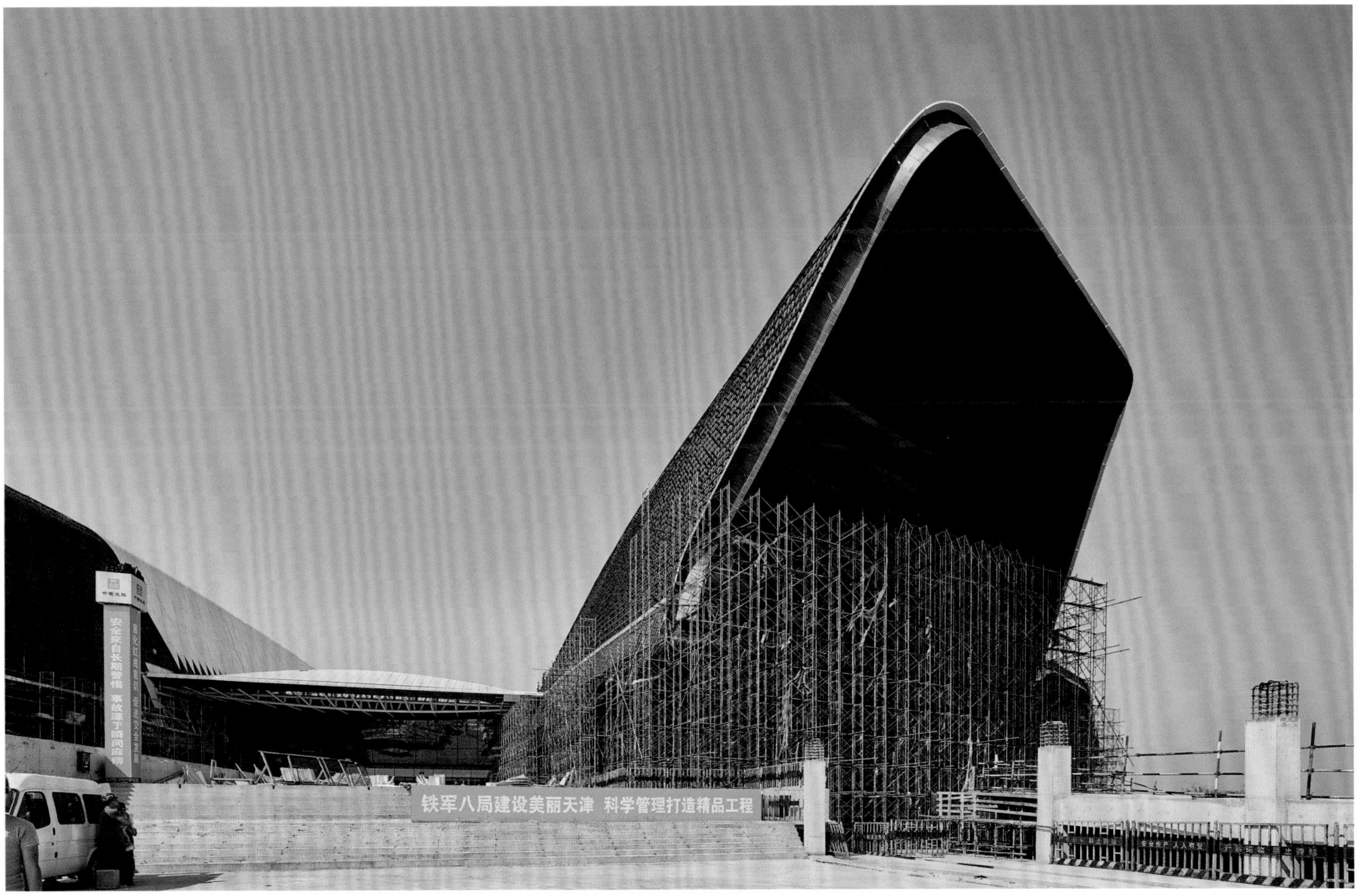
安全来自长期警惕 事故源于瞬间麻痹
铁军八局建设美丽天津 科学管理打造精品工程

建设过程中的国家海洋博物馆航拍

建设竣工的国家海洋博物馆航拍

国家海洋博物馆中国钢结构金奖现场评审会

四、机电设计

国家海洋博物馆建筑设计已荣获国家绿色建筑设计三星标识。机电设计充分利用可再生能源，广泛应用了太阳能热水系统、光伏发电系统、地源热泵系统、智能化管理系统等绿色技术措施。

太阳能热水系统。经多方案比较，结合天津地区丰富的太阳能资源及海博馆的地理位置，馆址处的年日照小时数超过 3000 h，水平面年太阳辐照量超过 5400 MJ/m^2，最终确定采用太阳能热水系统。结合建筑屋面将集热器布置于中央大厅锯齿形屋面上，首层设太阳能热水机房，辅助热源采用容积式电热水器。太阳能热水系统为公众自助餐厅、职工餐厅的厨房提供了生活热水，试运营后均达到绿色节能的预期效果。

光伏发电系统。采用三相并网（内部电网）并上网（市政电网）模式，系统内含 630 块多晶硅太阳电池组件，总装机功率为 160 kW, 为建筑内负载提供电力，做到多余电量上网，电量不足时由电网补充。光伏发电系统的输出并入首层变电站照明配电柜的低压母线上，为建筑内照明负载提供清洁的绿色电力。

地源热泵系统。从各种能源的技术、经济性、利用条件等角度，对空调冷热源方案进行了分析评估，本着绿色节能的原则，最终确定垂直埋管地源热泵系统。根据测试得到的井热响应数据报告，单位钻孔深度平均取热量为 38 W/m，放热量 72 W/m。本工程设计地源热泵换热井 1463 口，单口井深 120 m，埋管间距 5 m，占地面积约 36575 m^2。地源热泵系统的利用，大幅减少了一次能源的消耗，大幅减少了污染物的排放及大气污染，大幅减少了冷却塔容量，降低

了城市热岛效应，实现了“绿色”供能。

超大空间消防系统。国家海洋博物馆内部建筑空间形式丰富，有多处高大跨层空间且变化多样，机电专业在大空间消防设计中，采用多种方案，既满足消防安全要求，又与建筑展陈空间完美结合。首先，高大空间的消防灭火要求是普通自动喷水灭火系统不能满足的，因此结合建筑空间形式，设置了大空间智能灭火系统，并对其装置进行装饰，使之与周边建筑环境和谐统一。其次，对于展厅、中央大厅、入口大堂等，按照防火分区，设置机械排烟系统，机械排烟风按6次/h计算，排烟口距离最远点水平距离不超37.5m。最后，考虑高度超过12m的高大空间场所的建筑特点及在发生火灾时火源位置、类型、功率等因素的不确定性，采用高灵敏度的线型红外光束感烟探测器结合吸气式感烟探测器作为火灾参数探测器，进行火灾预报警及相关装置联动。

智能化管理系统。根据国家海洋博物馆展览特色，体现“互联网+”设计思维，设置了海博新媒体平台、掌上海博、智慧导览系统等管理系统平台，为观众及馆内工作人员提供不同层次的服务模式和多主线的服务功能体系。结合智能化的先进理念与系统产品，提供了一套技术架构成熟稳定、运行流畅、具有完备的扩展性和升级性能的设计方案，配置了如下各智能化系统。①公共安全系统：本馆的安全防范为一级风险等级，由视频安防监控系统、出入口控制系统及速通门系统、入侵报警系统、电子巡更系统、客流分析系统、安检门系统、可视对讲系统、安防系统集成管理系统、消防系统等组成。②建筑设备管理系统：由建筑设备监控系统、建筑能效监管系统等组成。③信息设施系统：由信息网络系统、综合布线系统、用户电话交换系统、移动通信室内信号覆盖系统、有线电视系统、公共广播系统、会议系统、信息导引及发布系统等组成。④智能化集成系统：实现信息、资源和任务共享，完成集中与分布相结合的监视、控制和综合管理。⑤机房工程系统：按照B级计算机房标准建设，满足计算机等各种微机电子设备和工作人员对温度、湿度、洁净度、电磁场强度等的要求。

1. 绿色三星建筑

国家海洋博物馆建筑设计已获得国家绿色建筑设计三星标识。设计中广泛应用了包括太阳能热水系统、光伏发电系统系统、地源热泵系统等技术。

（1）太阳能热水系统

充分利用太阳能资源，解决餐厅、厨房的生活热水问题，太阳能热水系统的集热器结合建筑锯齿形屋面的做法，安装于中央大厅屋顶。

（2）光伏发电系统

利用屋面架构的 30° 倾角顺坡铺设太阳能电池组件，实现与建筑屋面整体效果完美结合，解决了电池板的通风散热问题。同时，考虑

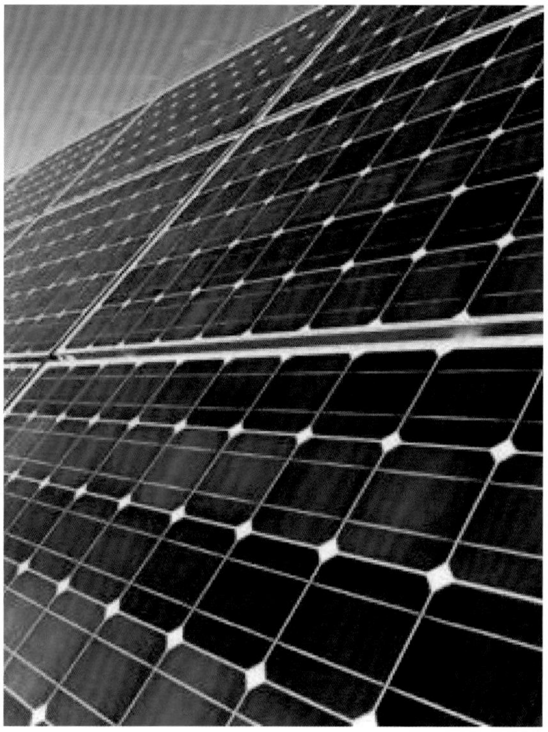

光伏太阳能板

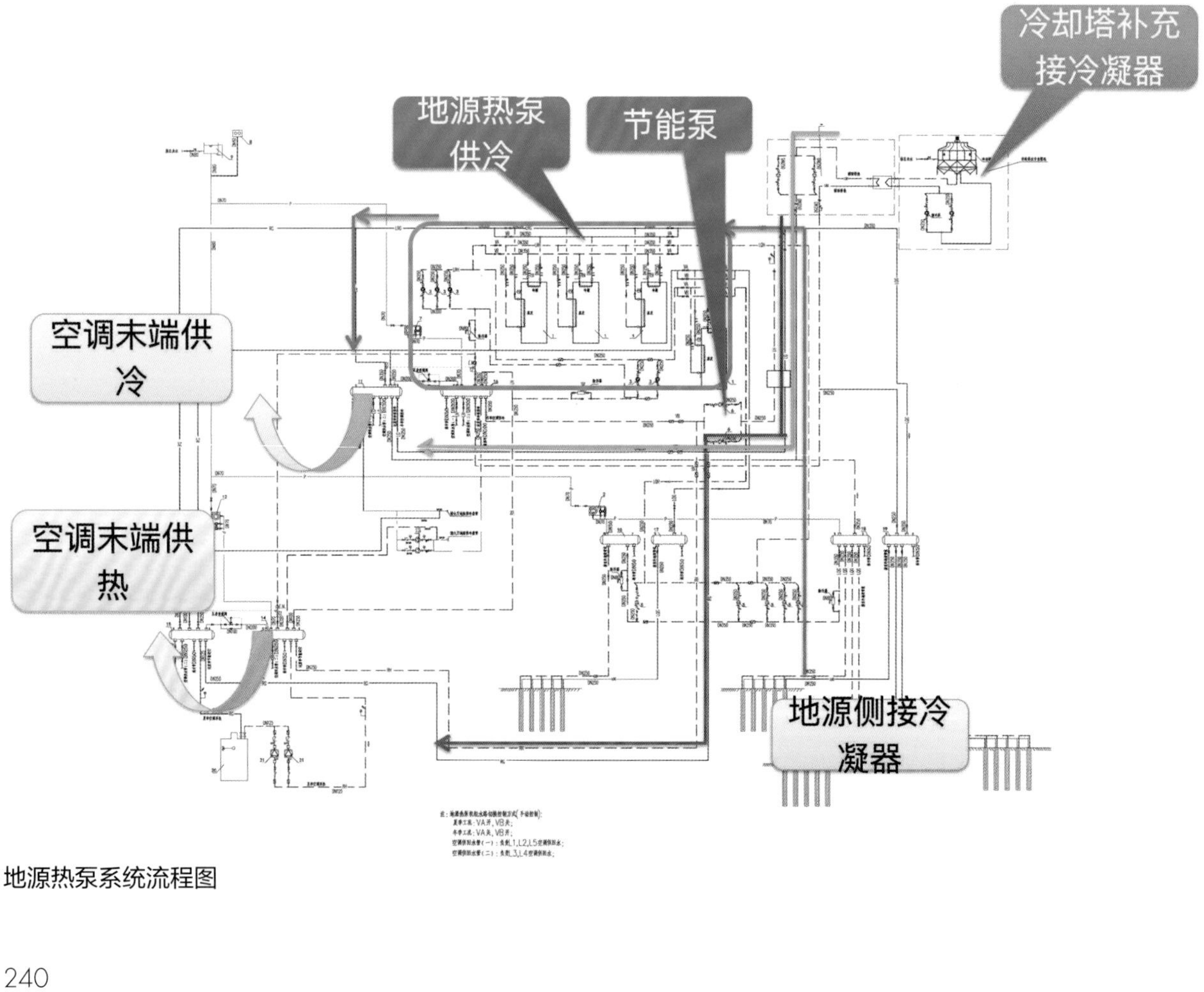

地源热泵系统流程图

北方的气候条件，为保证系统发电效率，在屋面设置多处自来水冲洗点，定期清洁板面上的尘埃。

（3）地源热泵系统

夏季：由地源热泵机组供冷，地源侧高温水及锅炉出水作为系统再热热源。为保证地源侧取放热平衡，设置一台冷却塔，根据地源侧回水温度确定冷却塔是否开启，作为地源侧散热的补充。

冬季：由地源热泵机组供热，地源侧低温水作为系统冬季热源；根据冬季冷负荷需求选取节能泵。同时以锅炉作为冬季地源热泵供热系统的补充，当系统初期运行地源侧工况不稳定时，可开启锅炉供热。

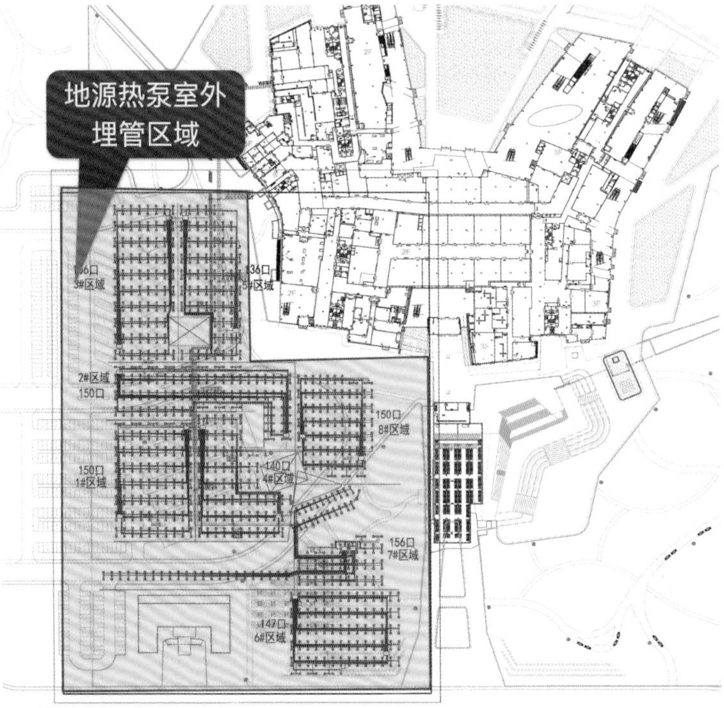

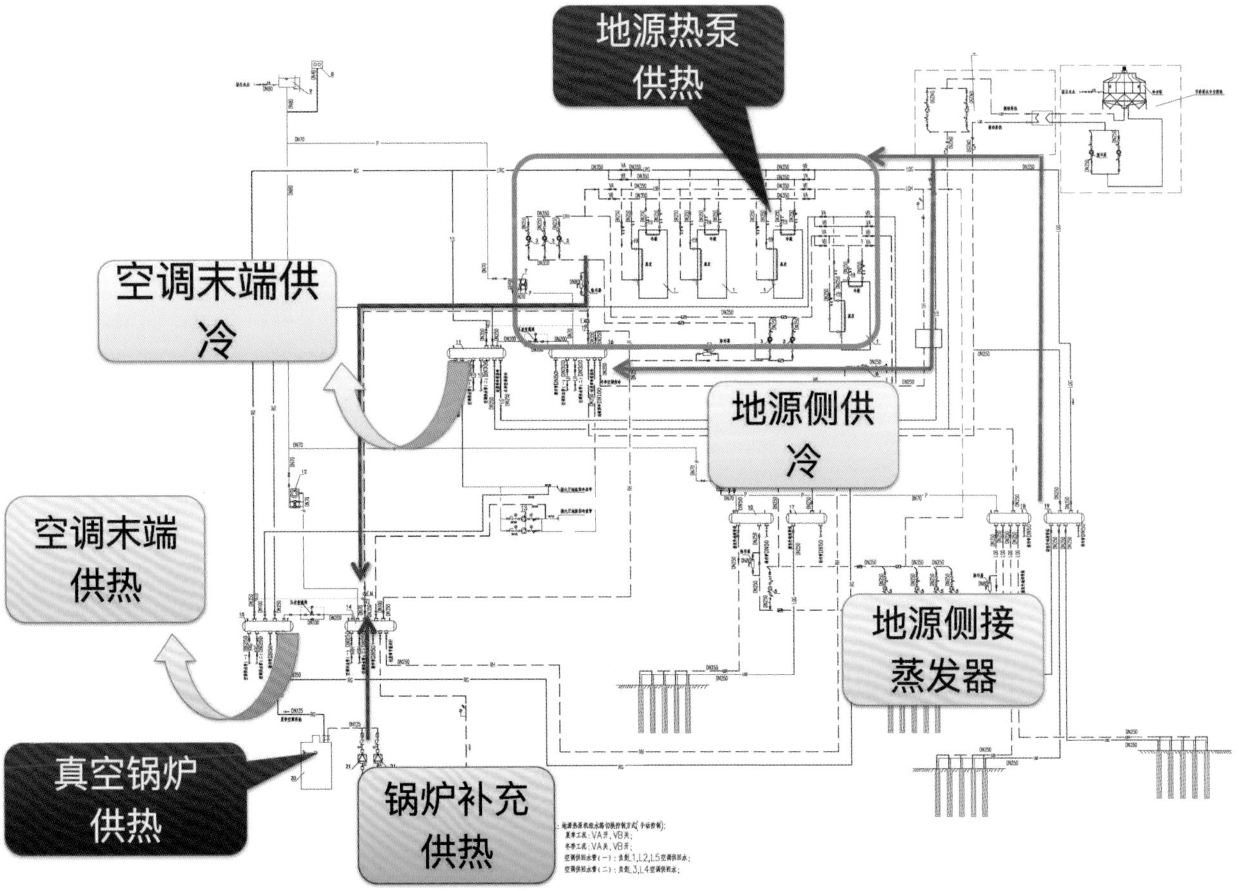

地源热泵系统流程图

2. 大空间消防设计对策

（1）大空间智能灭火系统

海博馆共享空间高度超高，一般喷淋系统无法满足发生火灾时的灭火要求，因此采用了大空间智能灭火系统，同时考虑空间美观问题，对大空间智能灭火装置进行装饰。

（2）大空间火灾自动报警装置

采用两种火灾参数探测器：线型红外光束感烟探测器、吸气式感烟探测器。

其报警时间早于传统探测设备，可以在火灾行程早期发现风险隐患，将发生火灾风险概率降到最小。

（3）大空间消防排烟设置情况

展厅、中央大厅、入口大堂设置机械排烟系统，结合建筑造型合理布置排烟口。

大空间内的消防系统设置（组图）

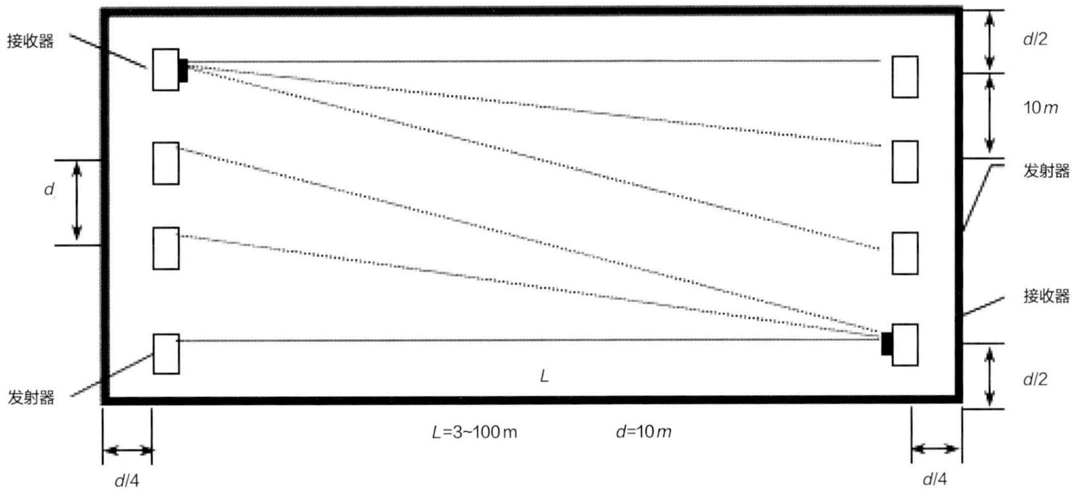
接收器
d/2
10m
发射器
d
接收器
发射器
L
d/2
L=3~100 m
d=10 m
d/4
d/4

3. 智能化系统设计

国家海洋博物馆设置了高质量、高标准、高要求的智能化系统。设计充分利用目前智能化领域中高端的设备和产品，考虑并结合博物馆行业特点，采用适用先进、适度超前、优化组合的成套技术体系，建成一座数字化、网络化、集成化的智能博物馆。

国家海洋博物馆智能化系统组成一览表

序号	1	2	3	4	5	6	7	8	9	10	11	12	13	14	15	16	17	18	19	20	21	22	23	24	25	26	27	28	29	30	31	32	33
系统组成	公共安全系统														建筑设备管理系统		信息设施系统									智能化集成系统		机房工程系统					
	安全技术防范系统						消防系统																										
智能化子系统	入侵报警系统	视频安防监控系统	出入口控制系统	电子巡更系统	安全检查系统	安全防范综合管理系统	火灾自动报警系统	火灾自动联动系统	消防应急广播系统	电气火灾监控系统	消防直通对讲电话系统	防火门及防火卷帘联动控制系统	消防电源监控系统	气体灭火系统	建筑设备监控系统	建筑能效监管系统	信息网络系统	综合布线系统	用户电话交换系统	移动通信室内信号覆盖系统	可视对讲系统	有线电视系统	公共广播系统	会议系统	信息导引及发布系统	智能化信息集成系统	智能照明控制系统	通信及有线电视接入机房	信息网络机房	消防控制室	安防监控中心	智能化设备间	机房综合管理系统

智能化系统（组图）

（1）公共安全系统

安防系统形成了由监视区（外广场、办公区出入口、主入口大厅）、防护区（允许公众出入的展厅）、禁区（储存藏品的库房、保险柜、修复室和其他不允许公众出入的区域）组成的纵深防护体系。

（2）建筑设备管理系统

建筑设备采用集成式控制，以实现节能、自动运行、控量等能源消耗情况的分项监测及计量，以及后台分析。主机采用双主机虚拟机运行方式，确保系统长期工作的稳定可靠。

现场智能控制面板具有场景现场记忆功能，以便于现场临时修改场景控制以适应不同场合的需要。

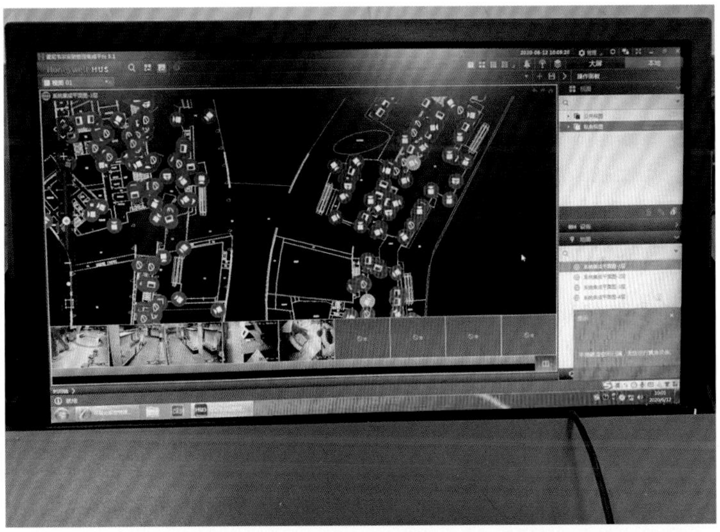

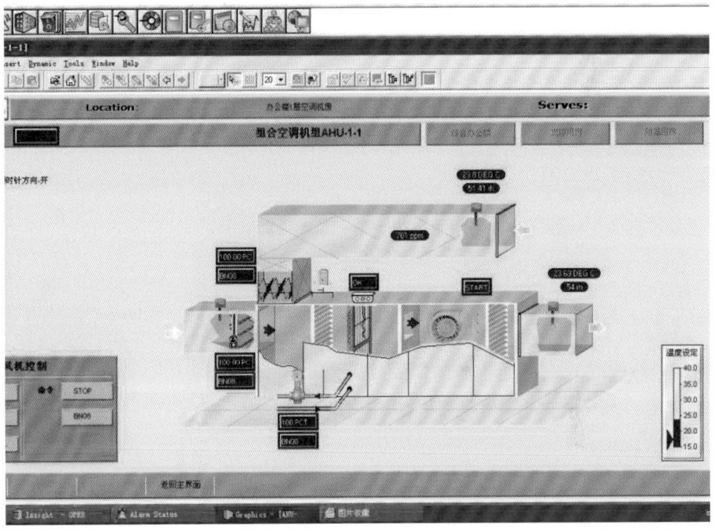

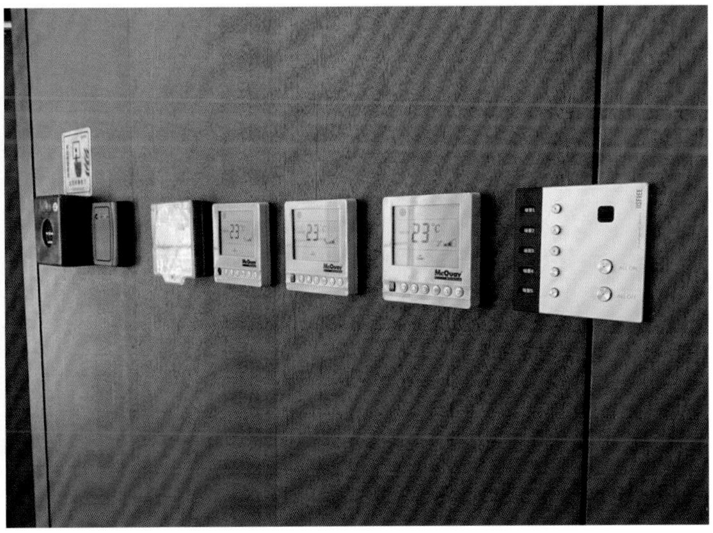

智能洗手间导引系统
男卫
女卫
有人
无人
Ask
中国海洋博物馆
智慧
导视系统
Intelligent
Guidance System
国家海洋博物馆

智能化系统 2（组图）

（3）信息设施系统

网络系统由内部综合业务网、外部应用服务网及安防专网共三套网络组成。均采用双核心、双链路、千兆接入、万兆上行的体系架构设计。

综合布线系统采用低烟无卤 6 类或 6A 结构化布线标准，千兆到桌面，主干万兆光纤上传，覆盖各办公区域、会议区、布展区域及藏品库区等。

（4）机房工程

按照 B 级计算机机房建设。

五、景观设计

随着城市建设的快速发展和人民生活水平的提升，人们对文化的追求也有着越来越高的标准。国家海洋博物馆是展示我国海洋自然及海洋人文的博物馆。国家海洋博物馆勾勒出中华文明有 2000 年历史的陆地黄色文明和海洋蓝色文明交相辉映的人文和自然景象。在这座重要的文化地标项目的景观设计中，以“馆园融合”为出发点，在陆地与海洋景观文化交织、外展与内展场域互动、绿色建筑景观措施等方面进行了全新的探索和体验。

国家海洋博物馆俯视图

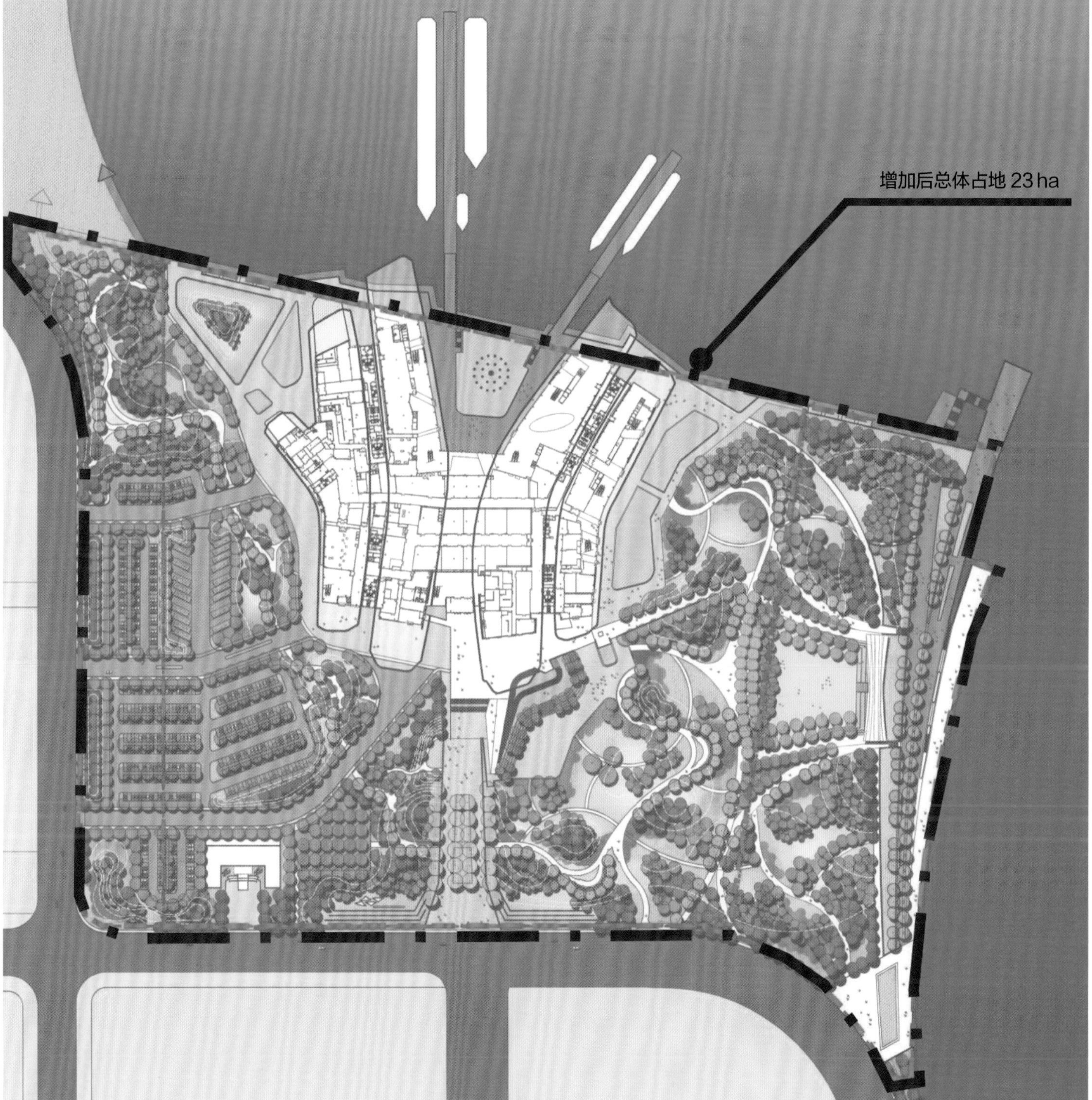
增加后总体占地 23 ha

1. 建筑与景观空间的融合——五大主题空间的融合

博物馆的景观设计是将景观、建筑、展陈、整体营造融为一体，按照从建筑到景观，由内到外，由陆到海，层层递进，反复融合来完成设计。

在环境空间一体化的大格局下，国家海洋博物馆形成了五大主题空间，即：入口广场空间、海博公园空间、滨水观景空间、滨水展示空间、绿色停车空间。五大空间各具功能特色，相互补充，相互交融，构成了国家海洋博物馆整体的景观空间布局。

（1）入口广场空间

入口广场空间纵深为 200 m，入口宽度为 60 m，通过入口大坡道与台阶的结合以及绿植形态烘托国家海洋博物馆建筑的整体形象与气势。

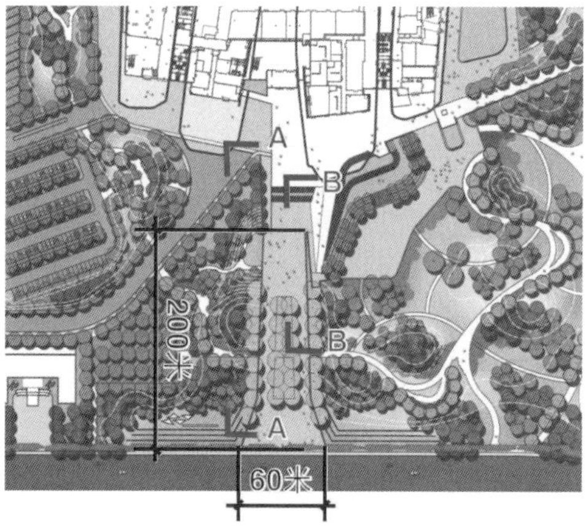

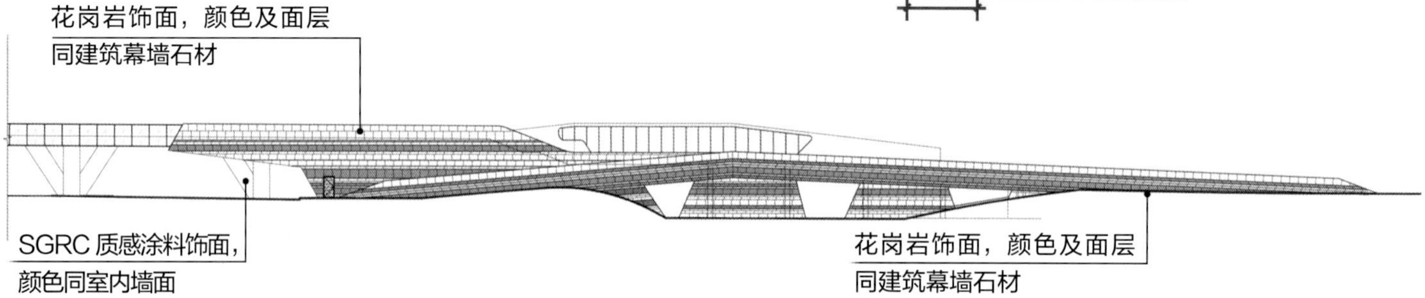

A—A 剖面

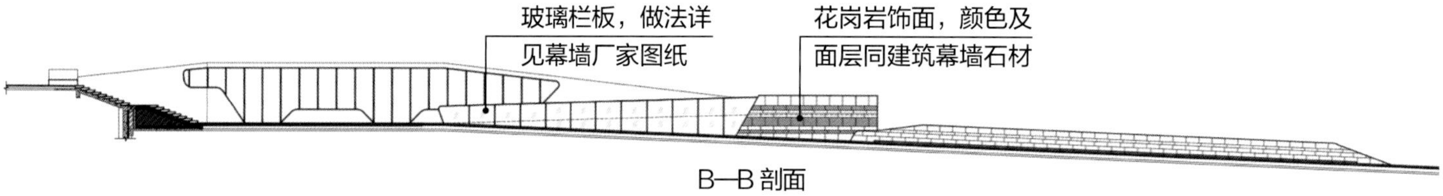

B—B 剖面

入口东侧公园广场的空间形态设计，与动态的建筑语言相得益彰，并运用了乔木种植与台地结合的方式形成了天然的绿荫看台，为海博馆举办室外活动提供了良好的空间体验。

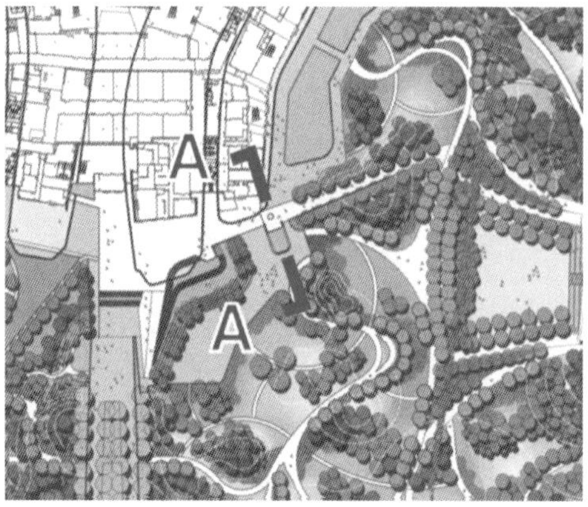

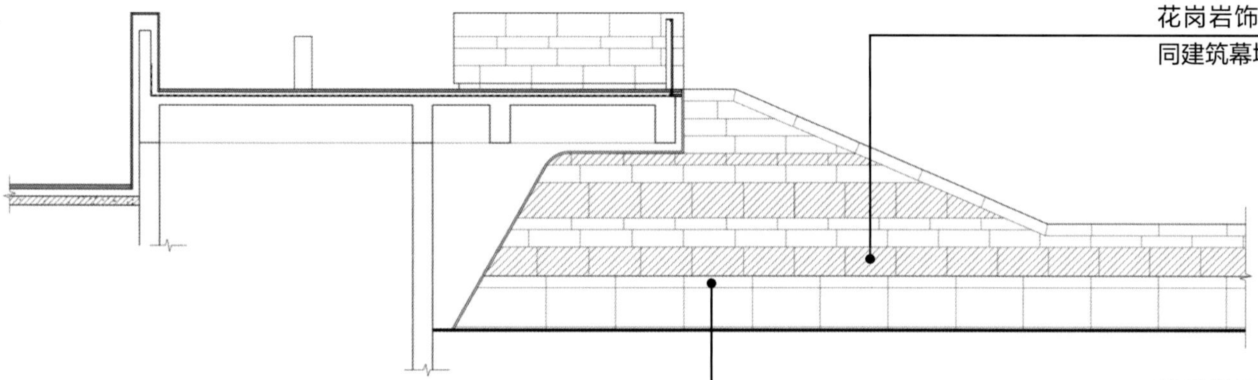

A—A 剖面

（2）海博公园空间

在空间形态设计上采用绿色生态流线的形式，体现海洋柔美的自然形态。绿植空间体系上运用色彩以及季相的变化给人以大绿生态的宜人空间感受。建筑的东西两侧运用了大面积的静水面设计，从多维度展现出建筑与水的互动与变化，并结合室外展陈的设计，让整个公园彰显外展与内展的统一。

在景观设计中运用“台田堆坡”的设计理念打造出多重绿化微地形空间，将植物与建筑融为一体。同时，通过多重植被的丰富配置、和谐共生，打造四季常绿三季观花的景象，创造富有生命气息的生态环境。

海博公园绿化及其植被分布（组图）

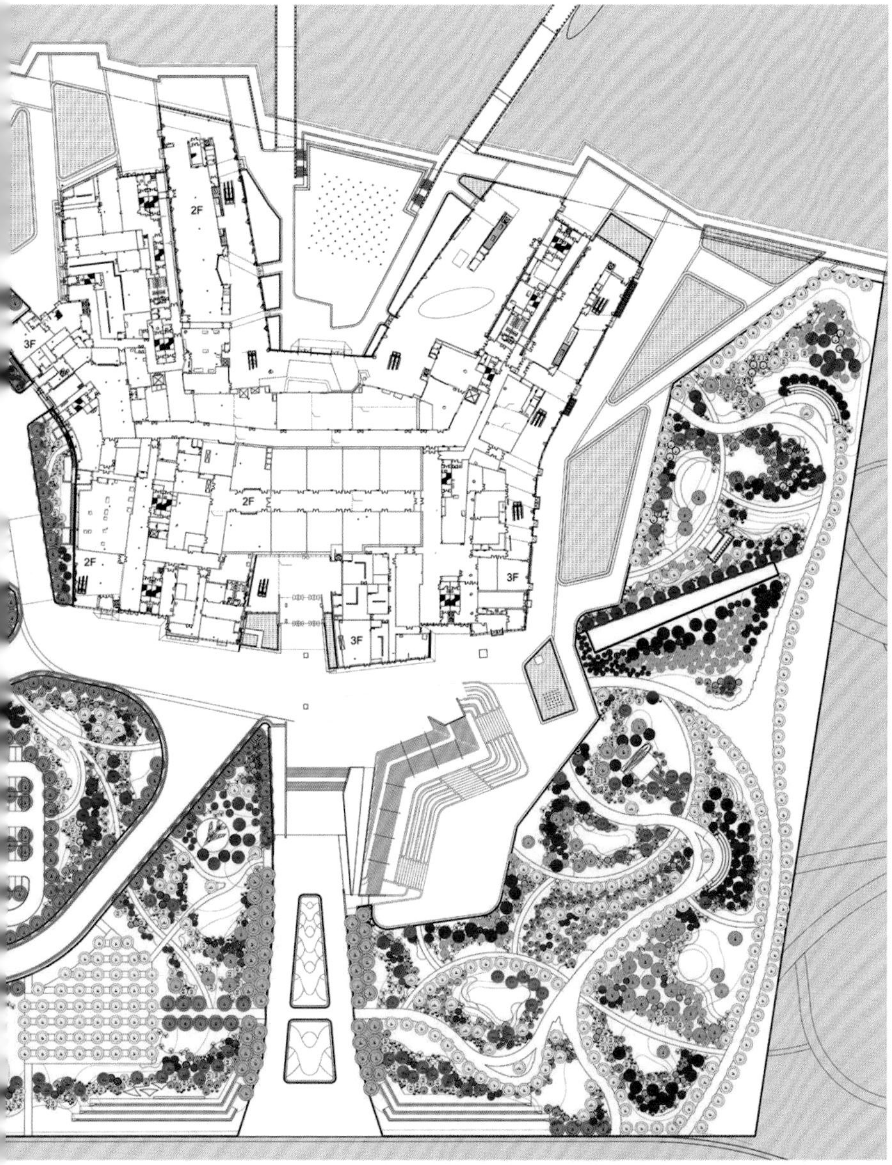
2F
3F
2F
2F
3F
3F

国槐
法桐
丛生元宝枫
丛生蒙古栎
栾树
白蜡
旱柳
垂柳
玉兰
白皮松
合欢
桑树
构树
雪松
碧桃
山杏
云杉
油松
山桃
山楂
樱花
紫叶李
西府海棠
绚丽海棠

（3）滨水观景空间

滨水观景空间运用向海面出挑的平台，使公园与海完美契合；公园东北角的观景平台与远景的妈祖像遥相呼应，从视觉上形成了海博馆区块与滨海旅游区景点的互动；从流线组织上也与南湾亲水慢行系统形成串接。

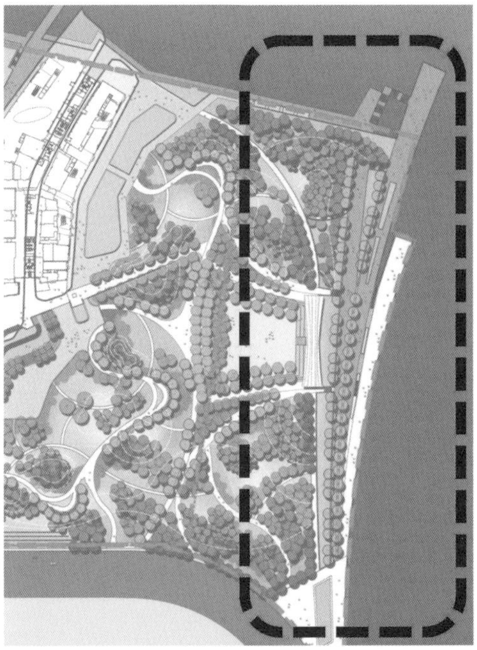

（4）滨水展示空间

滨水展示区由建筑北侧水景平台和伸出海面的两条栈桥组成。西侧栈桥边停泊着从海军退役的 752 导弹护卫艇，是海上展场的重要展示和爱国主义教育的重要展项。东侧栈桥为环南湾旅游船提供停靠和补给服务。东西向的滨水慢行系统与水上栈桥有机交融。

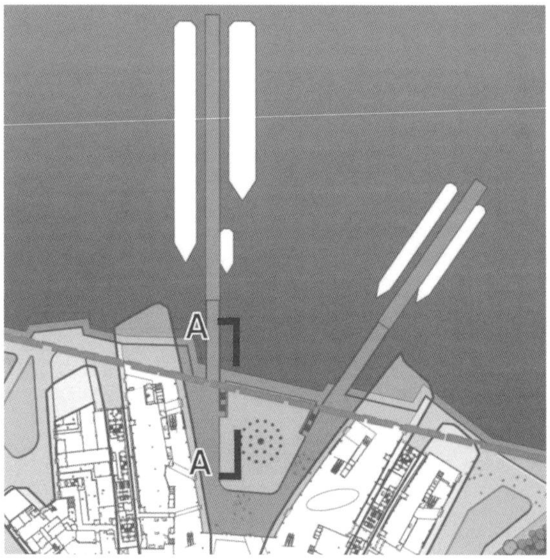

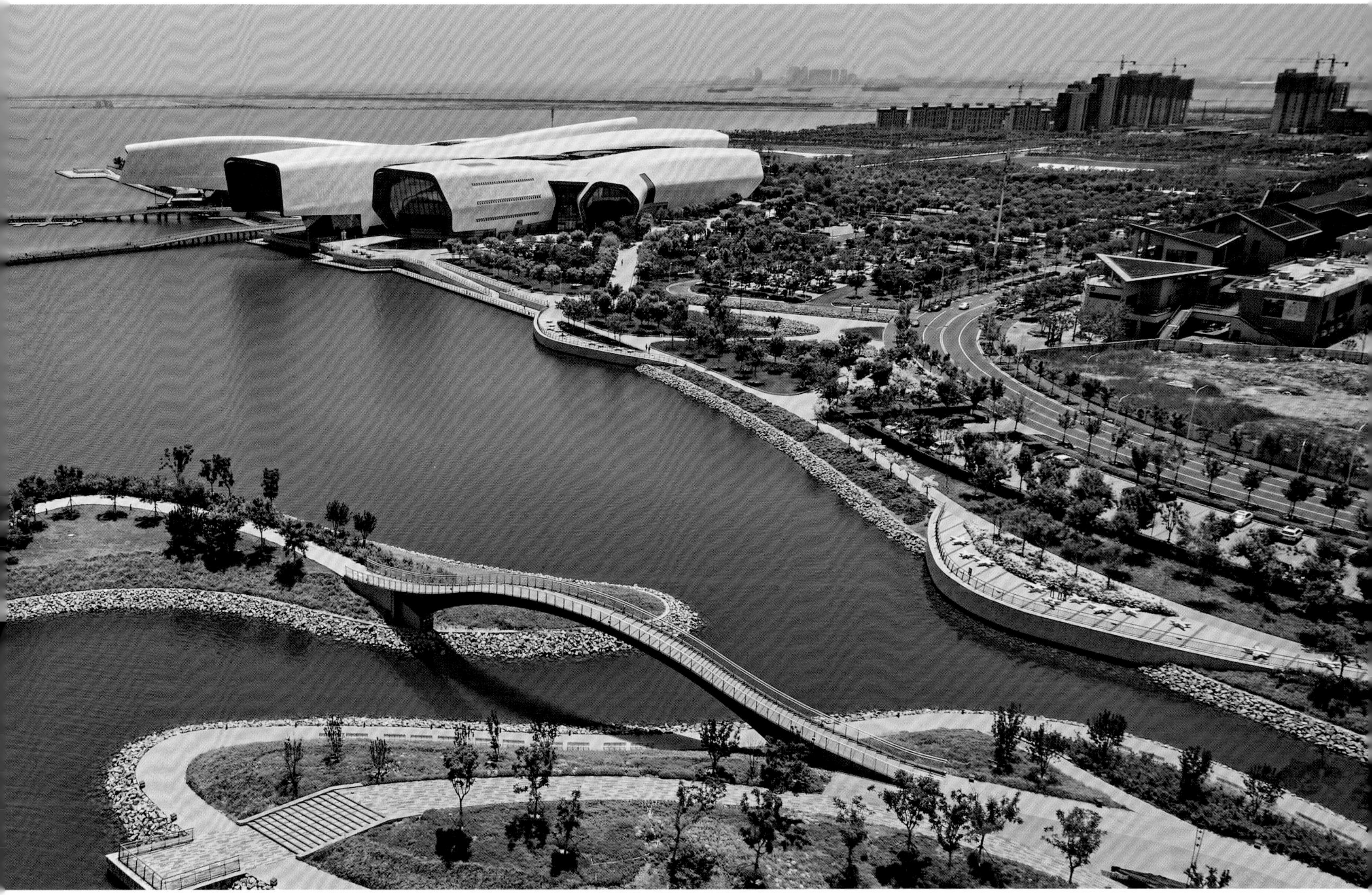

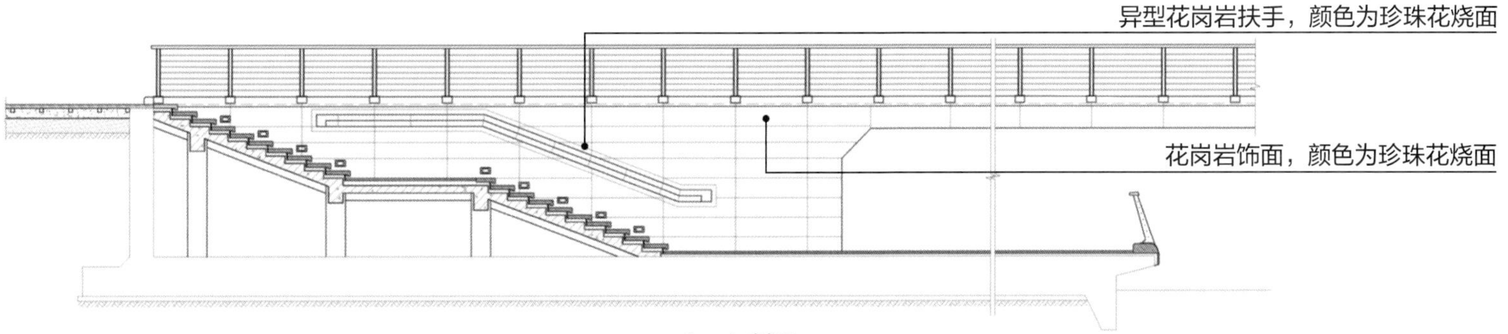
异型花岗岩扶手，颜色为珍珠花烧面
花岗岩饰面，颜色为珍珠花烧面

A—A 剖面

水景平台面向大海采用叠水系统装置，极大丰富了其观赏性和互动性。

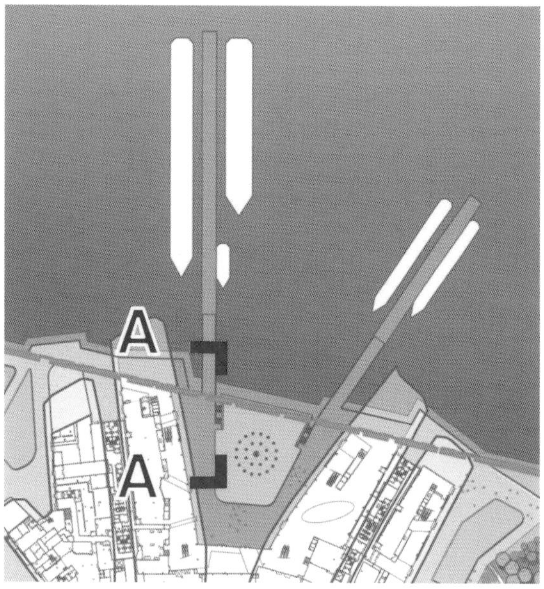

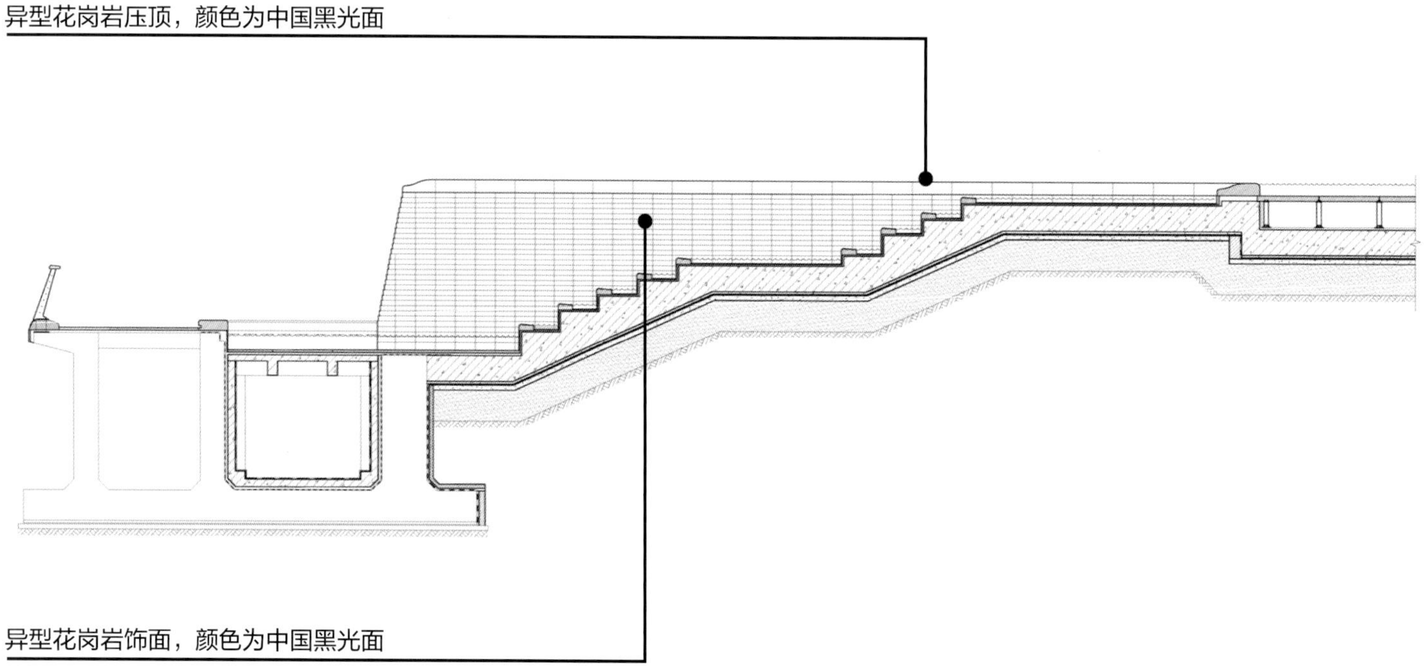

A—A 剖面

1）陆中有水。环绕建筑周边共有三片浅水池（东面 2145 m^2、中部 2643 m^2、西部 1515 m^2），水深 20 cm；浅水池池底通过艺术设计铺装三种色系马赛克，并形成斑斓的水纹图形，与海洋主题紧密契合，倒映在建筑玻璃幕墙上形成浑然一体的景观效果。

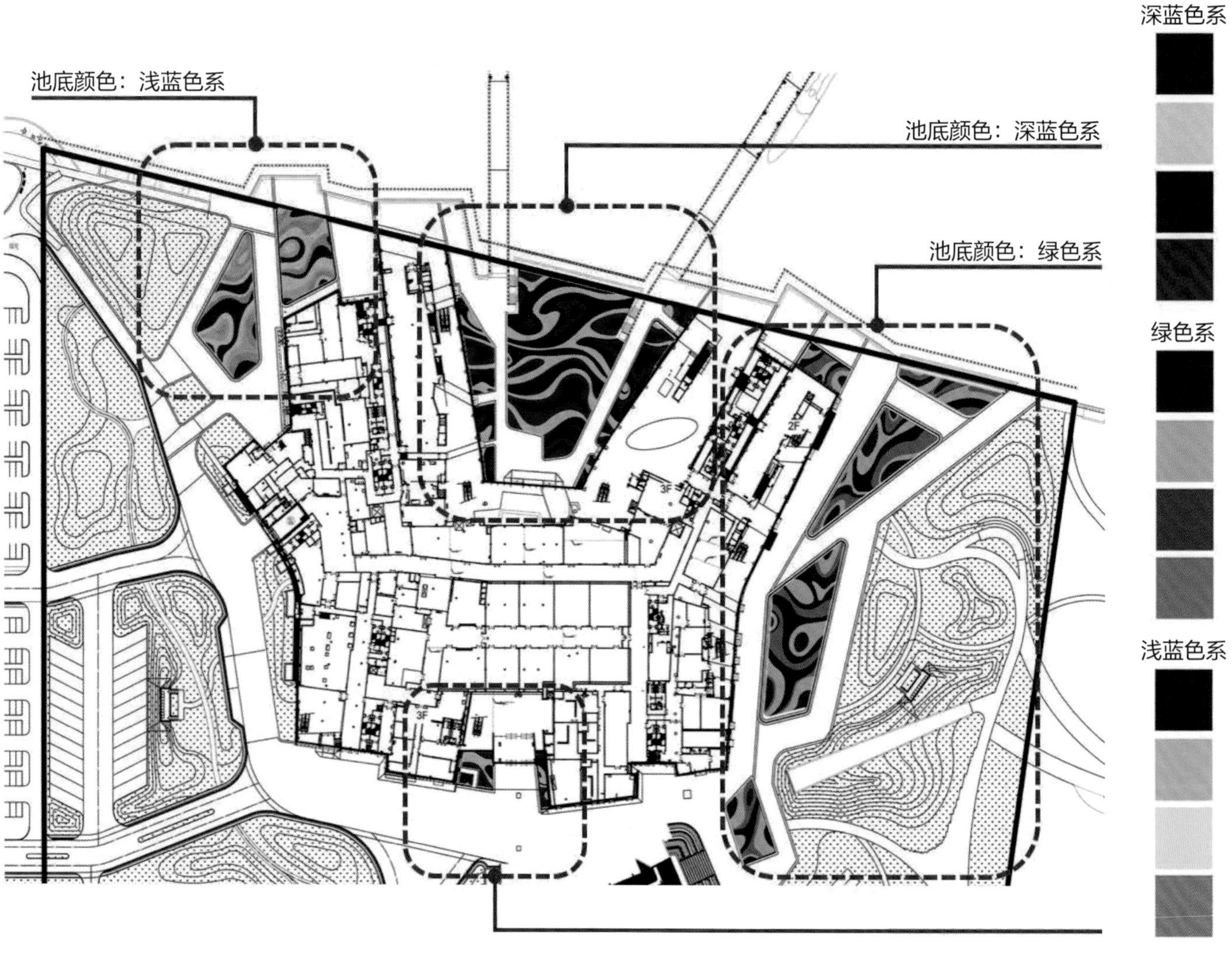

2）水中有路。建筑南侧设有两条栈桥（长度分别为 227 m、150 m，宽度为 8 m）。此区域室外地板铺装采取拼接龙骨后浇混凝土灌注的方式，加固基础以保证面层的牢固度要求。

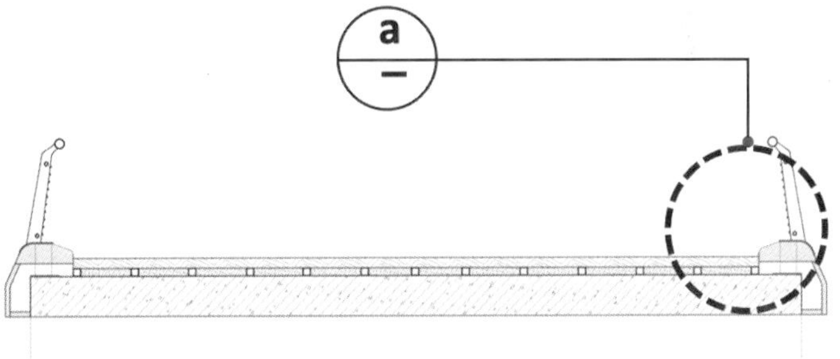

栈桥栏板剖面

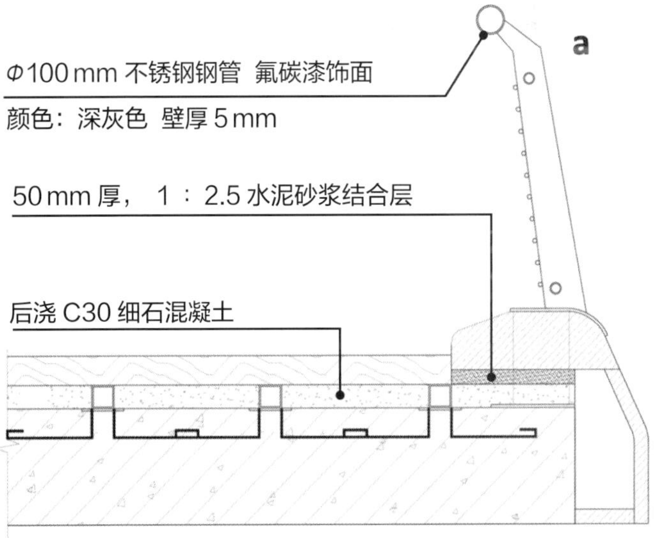

（5）绿色停车空间

在西侧停车空间设计了绿色停车系统，使每一个车位都能被绿茵所覆盖，通过地形高差设计让观众在停车区域形成沉浸式的体验。

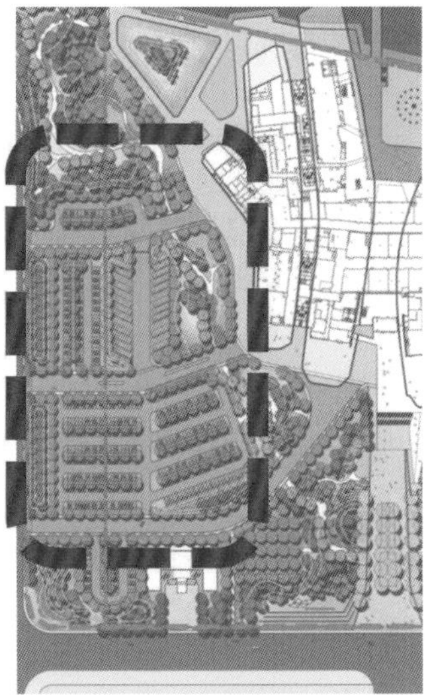

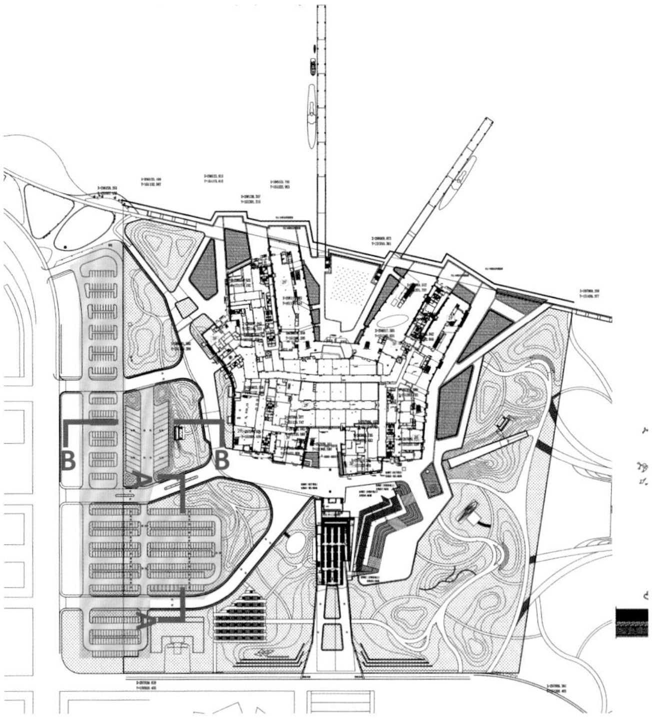

停车位平面图

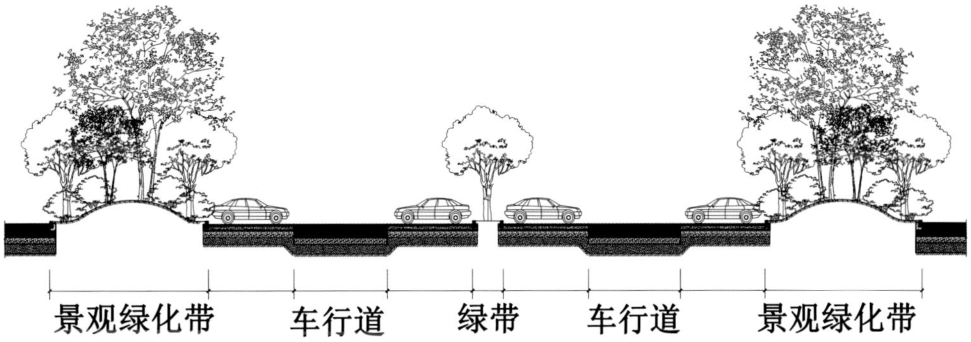

A—A 小型汽车停车位剖面图

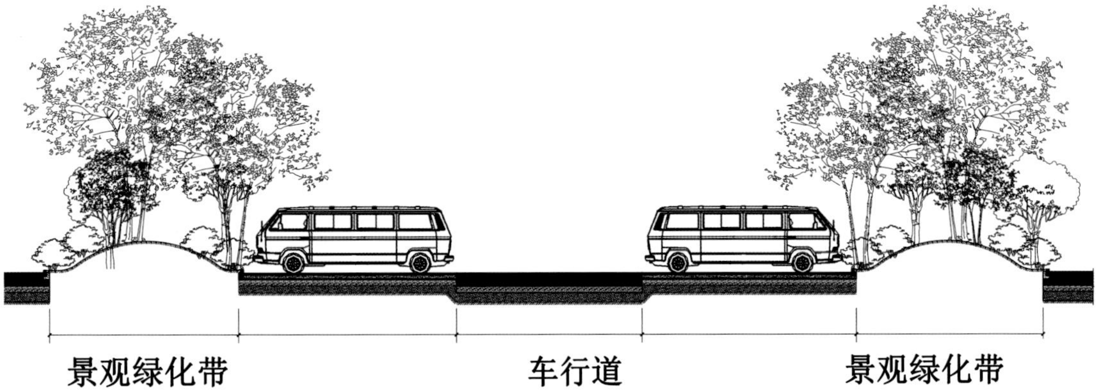

B—B 大型汽车停车位剖面图

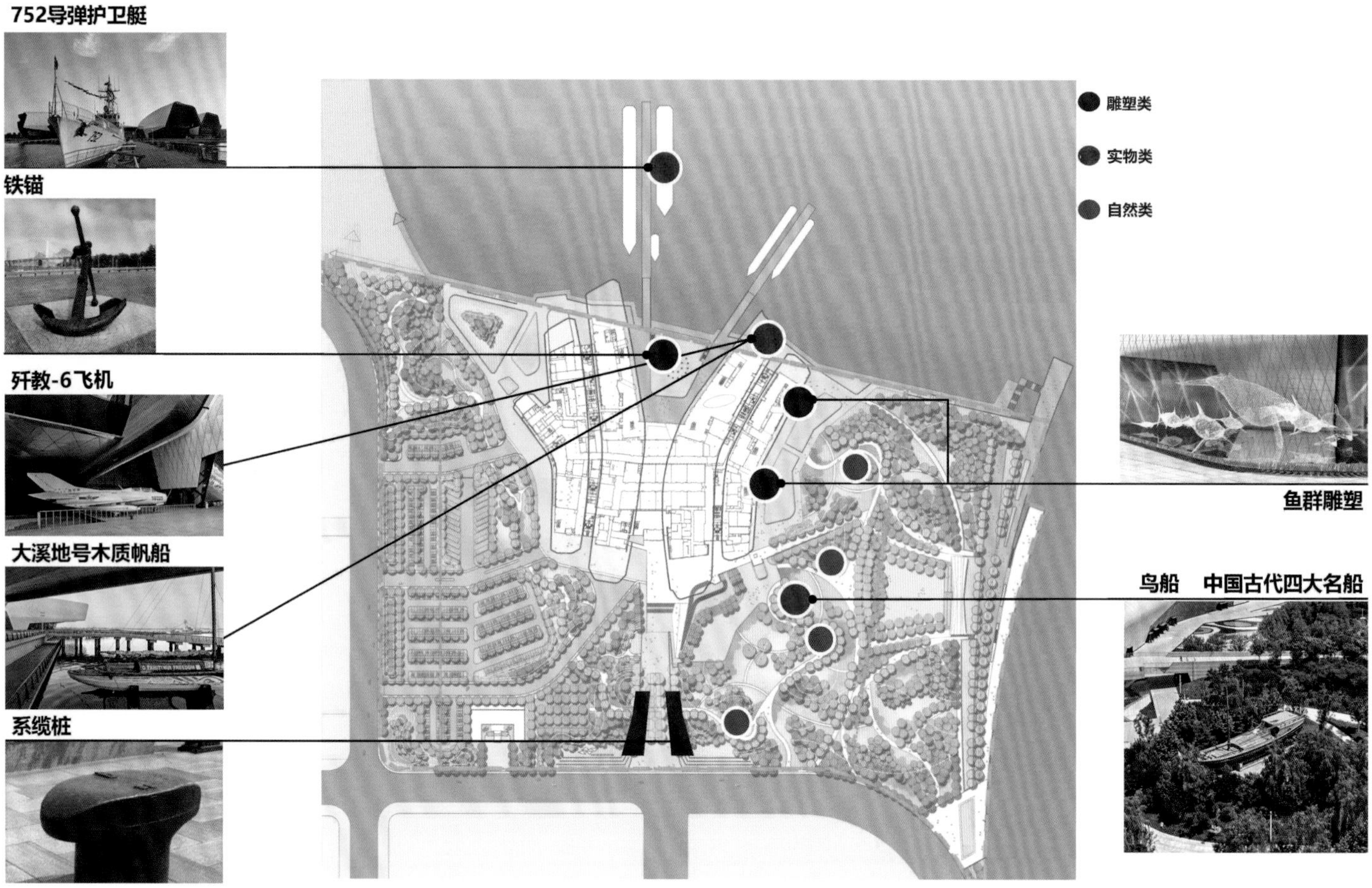

2. 外展与内展场所一体化

海洋博物馆不仅在 8 万 m^2 的馆内展示海洋自然和海洋人文的各种展项，在馆外的海博公园及海面上，均布置有相关展项。现有库存展品 5 万余件（套），上展展品近 8000 件（套）。展项归纳起来有三类：实物类、雕塑类、自然类。展项陈列场所的陆地部分与水域部分是水陆互动、内外互动并有启发性的，鼓励观众去探索与发现。

雕塑类——铁锚。在博物馆北侧中央水景中矗立一尊铁锚雕塑。该铁锚原形为 1866 年成立于福建福州的我国第一座被誉为中国海军建设摇篮的船政学堂。本雕塑寓意我国海军发展的时代特色，传承了一份与海洋相关的国家历史记忆。

雕塑类——系缆桩。在南入口坡道两侧共 12 个系缆桩雕塑，形成序列感和引导感，灯光效果使之形象与功能相得益彰。

雕塑类——鱼群。在位于博物馆东侧约 2000 m^2 水池上设计了一组跃向海面的鱼群雕塑，既为景观增添了活跃的海洋元素，也与建筑内展厅形成了内外互动之势。

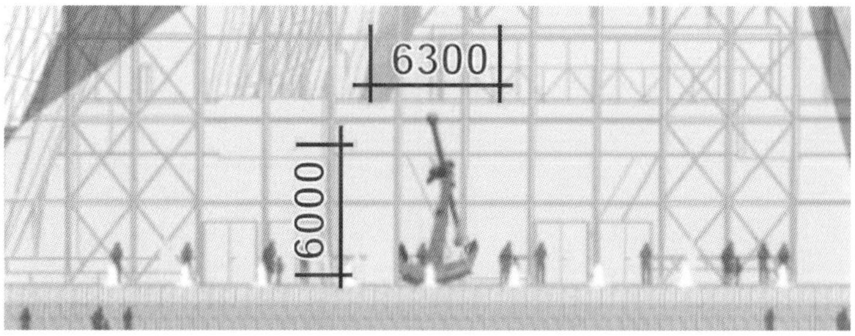

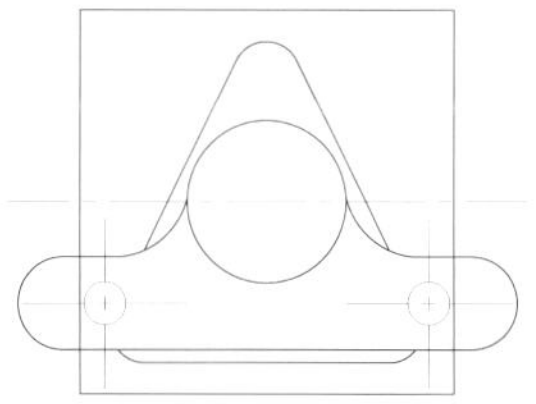

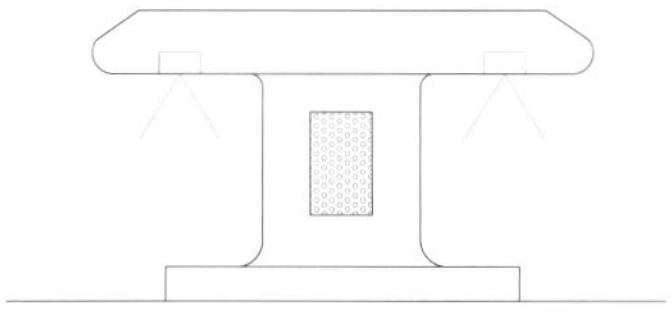

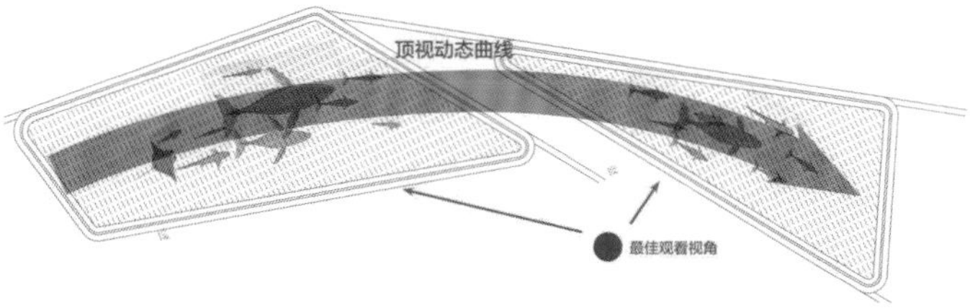

自然类——叠层石。叠层石是一种“准化石”，为亿年前退海为陆的见证，它象征地块成长的记忆。

实物类——752 导弹护卫艇。该艇于 1991 年下水，为中国海军第二代导弹快艇，由上海求新造船厂生产， 2014 年 8 月退役 。

实物类——大溪地号木质帆船。该船是波利尼西亚人建造的传统木质帆船，主体船长 15.3 m 浮架长 14.5 m，作为珍贵的历史证物展出。

实物类——乌船。该船修复后重约 60 t，船长约 25 m，最大处宽度约为 5.4 m，船高为 5.5 m，为典型浙江“乌船”，是中国四大名船之一。

实物类——歼教 -6 型超音速喷气式歼击教练机。该机由沈阳飞机工业集团设计制造，由南海舰队航空兵部队捐赠。

自然类展品（组图）

实物类展品（组图）

六、标识设计

导向标识系统作为连接访客与建筑环境间的桥梁，其本身也是整体环境的一部分。国家海洋博物馆的整体建筑设计采用了隐喻的手法将海洋船舶和海洋生物进行结合，其建筑本身可谓是独一无二的，视觉识别度极高。所以其标识系统的设计风格应与整体建筑与精装高度统一，做到“与建筑共生”。在整体的外观与材质选择上，充分融合了建筑及 精装的风格，甚至在木纹和喷涂的工艺上，更是直接使用了精装元素，最大幅度地做到醒目而不夺目，融合而不隐匿的效果。其次，作为一个海洋博物馆的导视系统，海洋元素的合理融入是必要的，但是还需要把握好尺度，不能过度表现，以免整体偏重艺术和卡通化，从而失去一个国家级博物馆的庄重感，所以设计只在细节处点缀了如锚型的箭头、象征

关于造型
ABOUT MODELING
主题呼应
从“海浪”的视觉印象获得造型灵感
与国家海洋博物馆展览主题呼应

标识形式分析（组图）

海浪的曲线、海洋的颜色等，使得标识系统的整体设计理念既符合海洋博物馆的属性，又不失国家级博物馆的形象。

国家海洋博物馆的坐落位置远离市中心，参观人群主要采用自驾、团体公共交通的方式。博物馆规划了 600 多个地上停车位，并设置了三个社会车辆入口，一个员工车辆入口。三个社会车辆入口又分为两个小型机动车入口和一个大型机动车入口。为了实现合理的人流导向，在还未进入园区处就设置了相应的导向标识，进行人车分流指引，利用四级导向的程序，一步一步将游客导引到相应的位置。国家海洋博物馆的建筑主体为 3 层 (局部为 4 层)，陈列展览内容围绕“海洋与人类”主题展开，分为“海洋人文”“海洋自然”两大部分。同时设有博物馆商店、餐厅、咖啡厅、 影院等公共服务设施。常设展厅的主题设置极具逻辑性，游客应尽量按照馆方提前设计的参观路径进行游览，才能比较系统地依据时间逻辑并且较全面地参观到博物馆的各类展品。这就要求导视系统能够给游客规划出一个合理的路径导向。同时对于一些较为隐蔽却又十分重要的位置，如公共卫生间、安全出口，设计采用了发光标识的处理方式。 另外，院内外还设置了若干的警示和提示类标识，从细微之处给游客以安全的提醒和帮助。

国家海洋博物馆的导视系统设计充分了考虑了建筑和精装的关系，并分别站在馆方和游客的角度，理解不同功能的使用需求，在此基础上审视标识的布点以及信息是否合理，尽量做到规划无盲区，导向无死角，隐患有警示，服务人性化，最终通过导向标识的语汇，向游客讲述这座宏伟的建筑所涵盖博大精深的空间信息。

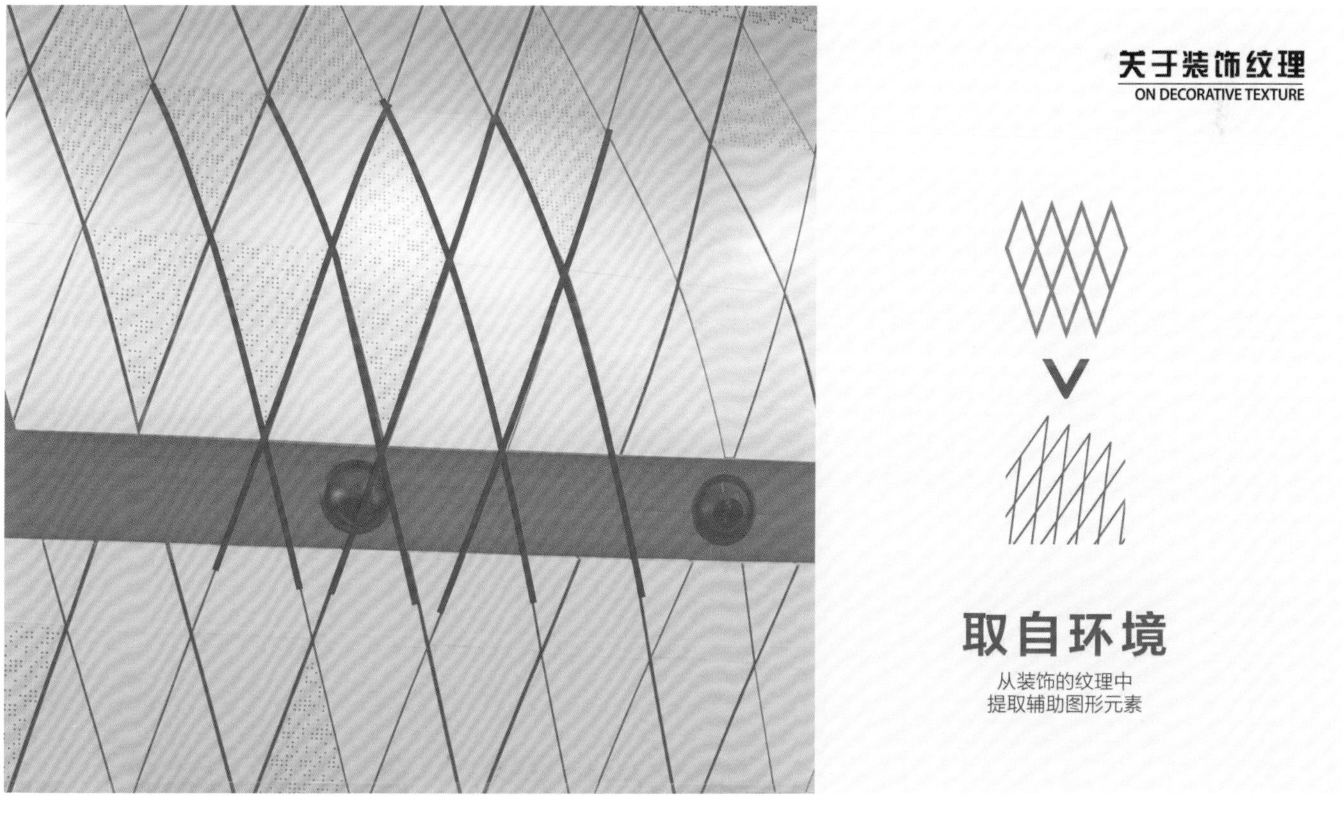
关于装饰纹理
ON DECORATIVE TEXTURE
取自环境
从装饰的纹理中
提取辅助图形元素

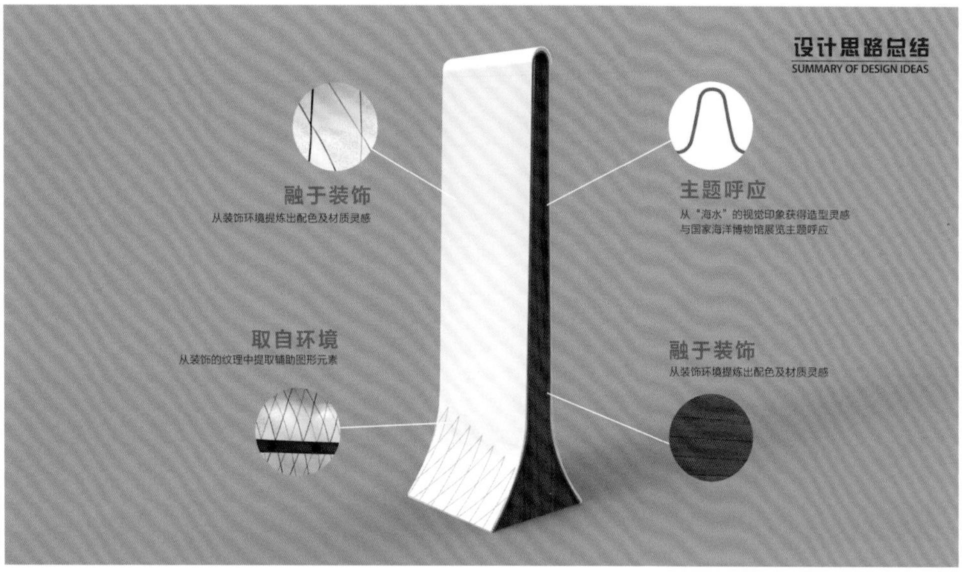
设计思路总结
SUMMARY OF DESIGN IDEAS
融于装饰
从装饰环境提炼出配色及材质灵感
主题呼应
从"海水"的视觉印象获得造型灵感
与国家海洋博物馆展览主题呼应
取自环境
从装饰的纹理中提取辅助图形元素
融于装饰
从装饰环境提炼出配色及材质灵感

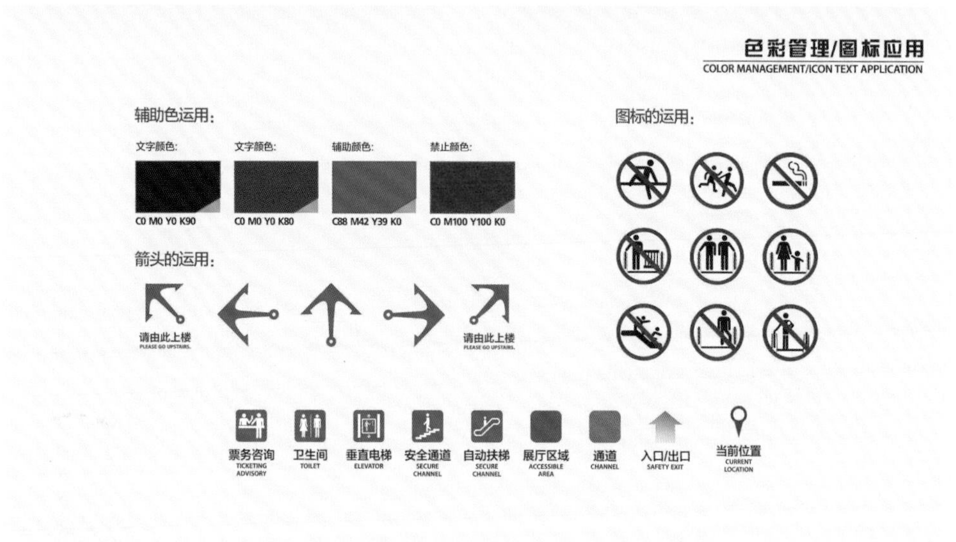
色彩管理/图标应用
COLOR MANAGEMENT/ICON TEXT APPLICATION
辅助色运用：
文字颜色：
文字颜色：
辅助颜色：
禁止颜色：
C0 M0 Y0 K90
C0 M0 Y0 K80
C88 M42 Y39 K0
C0 M100 Y100 K0
箭头的运用：
请由此上楼
PLEASE GO UPSTAIRS.
请由此上楼
PLEASE GO UPSTAIRS.
图标的运用：
票务咨询
TICKETING ADVISORY
卫生间
TOILET
垂直电梯
ELEVATOR
安全通道
SECURE CHANNEL
自动扶梯
SECURE CHANNEL
展厅区域
ACCESSIBLE AREA
通道
CHANNEL
入口/出口
SAFETY EXIT
当前位置
CURRENT LOCATION

文字应用
TEXT APPLICATION
中英文显示比例关系：
国家海洋博物馆
NATIONAL MARINE MUSEUM
中文主要字体：
方正兰亭中粗黑
国家海洋博物馆
英文主要字体一：
Myraid pro
NATIONAL MARINE MUSEUM
中文主要字体：
方正兰亭中黑
国家海洋博物馆
英文主要字体二：
Myraid light
NATIONAL MARINE MUSEUM
阿拉伯数字主要字体：
Univers
0123456789

导视设计案例（组图）

室外标识
Exterior Landscape Space
01
户外总导览图标识

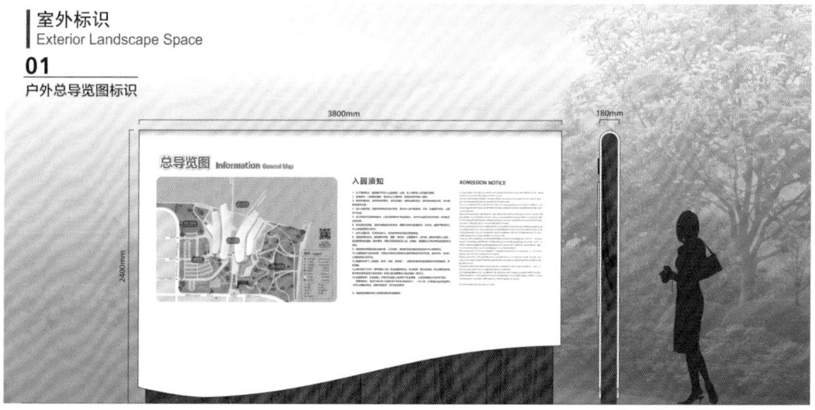

室外标识
Exterior Landscape Space
02
主干道指引标识

室外标识
Exterior Landscape Space
03
风向标指引标识

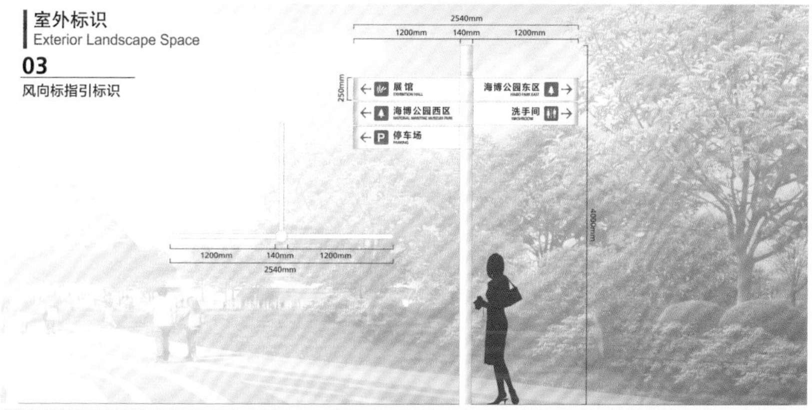

室内标识
Interior Signage
04
室内综合资讯标识

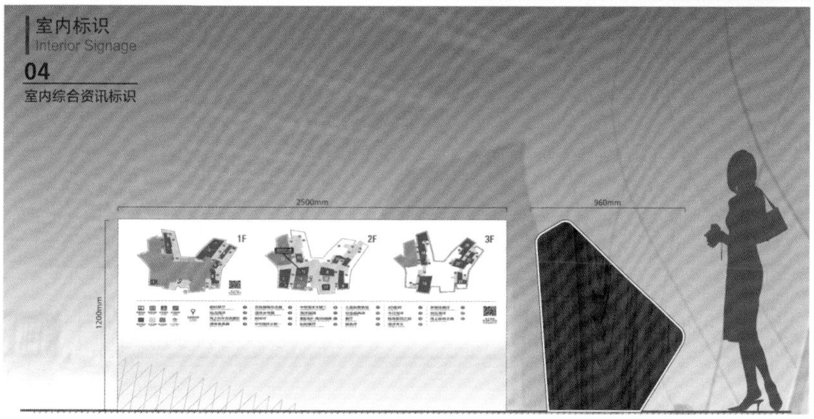

设计与建造实景对比（组图）

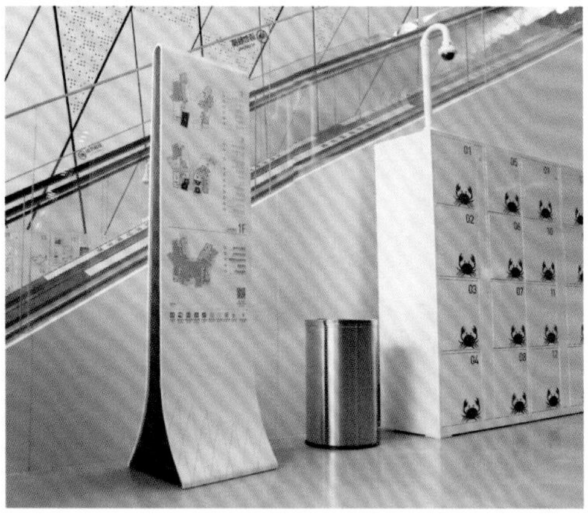

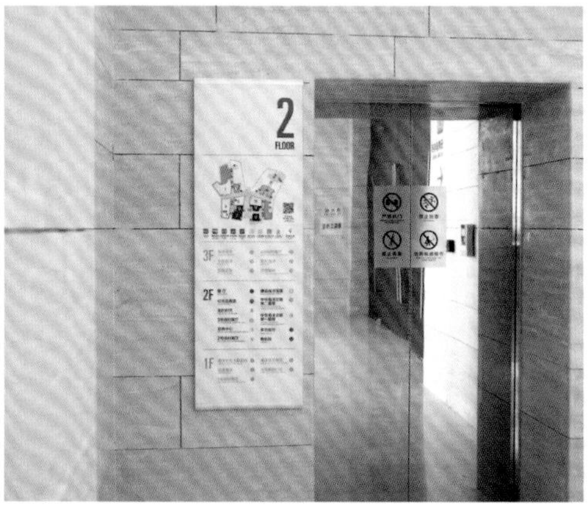

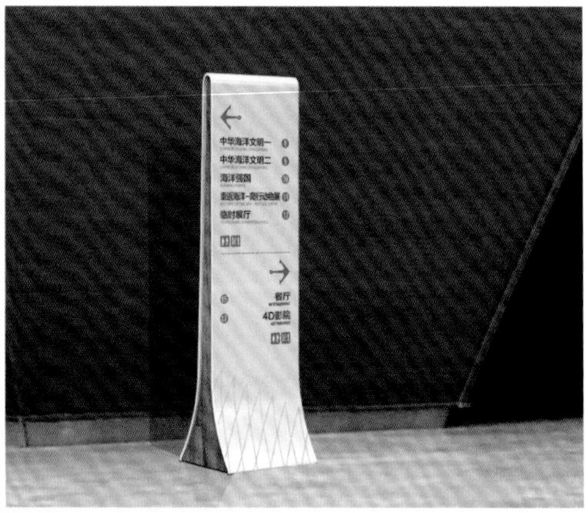

设计与建造实景对比（组图）

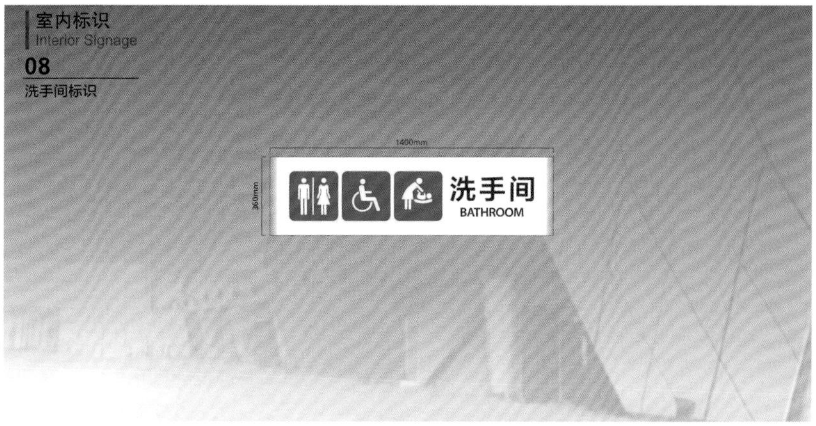
室内标识
Interior Signage
08
洗手间标识
1400mm
360mm
洗手间
BATHROOM

洗手间
TOILET

室内标识
Interior Signage
09
卫生间区域安全标识
300mm
300mm
400mm
请勿吸烟
NO SMOKING

禁止吸烟
NO SMOKING
饮水处

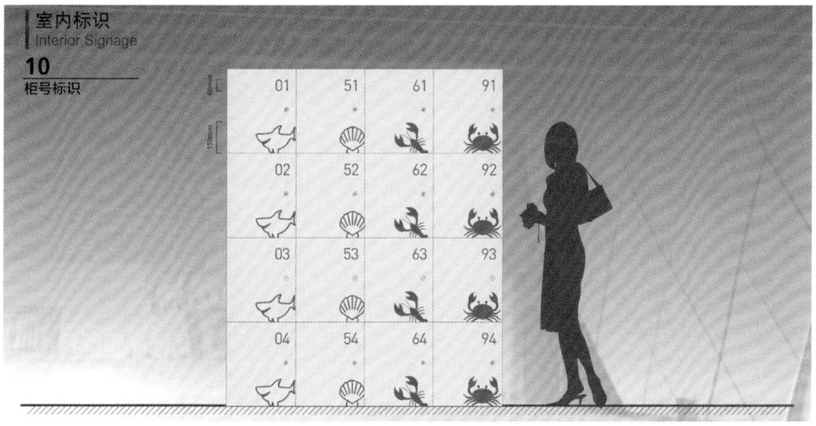
室内标识
Interior Signage
10
柜号标识
01 51 61 91
02 52 62 92
03 53 63 93
04 54 64 94

第四章　建筑月志

Chapter IV Monthly Records of the Construction

本章以技术与事件融合的记述方式，将时间轴与图片相结合，以工程实录的形式展现了自国家海洋博物馆奠基至竣工这一过程中，各方人员紧密配合、辛勤奉献的点点滴滴。它不仅准确反映了海博馆在设计与建造过程中那些值得记录的关键节点，也追踪设计人员于项目不同阶段为技术的革新与突破所留下的足迹。正是这样详实的工作，突显着大国工匠的伟大精神，全面而完整地展示了一个如此宏大且高完成度的建筑作品是如何通过各方人员紧密配合而实现的。

This chapter adopts a narrative style of integrating techologies and events, combining the time axis and pictures. In the form of engineering records, it presents the close cooperation and diligent work of people from different parties in the construction of the National Maritime Museum of China, from foundation laying to completion. The chapter not only accurately reflects the key points worth recording in the design and construction process of the museum, but also tracks the footprints left by designers for technological innovation and breakthrough at different stages of the project. It is such solid and meticulous work that highlights the great craftsmamship in a major power, comprehensively demonostrating how such a grand and highly accomplished architectural work is realized through close cooperation of professionals from all parties.

建设简述

国家海洋博物馆是一栋现代化、造型新颖的超大型公共建筑，建筑造形为多变的非线性曲面，主体门式桁架均为异形，建筑内部空间复杂，是由多种结构体系组合而成的铰接平衡体。面对建筑工程量大、技术难度高、施工环境复杂的建设要求和挑战，全体施工团队以精心施工的“工匠精神”，与设计团队密切配合，破解一道道技术难题，主要体现在以下四个方面。

1）建筑外围结构复杂，体量大。因造型新颖，极具未来气息，曲线形屋面多变，平面、立面均呈多曲弧面造型。整个外檐幕墙和外墙体系将近 $10\times10^4 m^2$，主要为 $6\times10^4 m^2$ 直立锁边铝镁锰金属板围护结构，上覆 3 mm 厚菱形铝合金穿孔饰板，建筑场馆端头及局部采用 $1.2\times10^4 m^2$ 玻璃幕墙，首层局部辅以 2000 m^2 石材幕墙。

2）门式钢桁架异形跨度大，建筑空间体量大：本工程共有 112 榀门式钢桁架，每个门式钢桁架的大小、造型、截面、安装角度均不一样。桁架圆管最大截面 Φ950 mm × 47 mm，桁架间斜撑连接杆件最大截面 Φ450 mm × 16 mm。需现场安装的钢构件近万件。单榀桁架最重为 93 t，最高为 34 m，最大宽度为 45 m。桁架构件倾斜设置，最大倾斜角度达 28.8° 。钢构件最大壁厚为 70 mm，单件最大吊重约 36 t。桁架体系并排设置，在空中相互连接，受力体系十分复杂，桁架底部采用万向抗震球铰支座与地面结构相连。

3）多结构体系组合的铰接平衡结构。国家海洋博物馆工程的异形钢结构体系由异形门式桁架体系 + 框架结构体系 + 异形悬挑桁架体系 + 桁架间斜向支撑体系 + 球铰支座组合构成铰接平衡结构体。

4）施工环境复杂。国家海洋博物馆坐落在渤海之滨的中新天津生态城南湾的海域吹填区，且北侧临海 55 m，悬挑钢结构跨在海面之上，为吹填未预压区域，是本工程的最大难点。

施工总包单位组织各方精兵强将，多方论证，有针对性地提出实施万向抗震球铰支座深化及施工综合技术、异形超高大跨度门式管桁架安装技术、跨海超 55 m 悬挑异形组合

2014 年 10 月 28 日 项目奠基

结构施工技术、异形复杂结构三维测量定位技术共四大关键施工技术，成功解决施工难点问题。

项目以上述技术为依托，并全程结合三维数字化技术相关软件，采用 4D 工期演示等技术内容对建筑外观造型、结构施工、构件定位等进行施工指导和监控。施工人员克服困难，攻破难题，夜以继日，最终交出一份高水平、高质量、高完成度的答卷，取得了优异的建设成果。

其中，异形超高大跨度门式管桁架框架结构体系关键技术，为国内领先水平，通过对关键技术进行研究，使施工过程中的质量、安全、工期及经济等得到合理控制并取得了良好效益。在质量上，通过以上关键技术的实施，保证了工程一次成优，顺利通过“海河杯”优质结构认定验收，于 2017 年 4 月顺利通过“天津市钢结构金奖”验收，于 2017 年 6 月顺利通过“中国钢结构金奖”验收，摘得中国建筑钢结构行业工程质量的最高荣誉奖，为后期“鲁班奖”验收提供质量保证。在安全上，上述技术的实施减少了安全风险隐患，并被评为天津市安全文明工地、观摩工地。在工期上，通过技术创新降低了施工难度，提高了施工效率，加快了施工进度，提前完成了主体验收目标。

2014 年 10 月 28 日—2015 年 3 月 30 日，桩基施工

2014 年 **2015 年** **3 月**

2015 年 3 月 31 日，主体开始施工，土方开挖

015 年 4 月，土方开挖完成

2015 年 4 月 20 日，地基验收

4 月

5 月

015 年 4 月，基础垫层施工完成

2015 年 5 月，基础地梁钢筋绑扎

2015 年 5 月，基础混凝土浇筑完成

2015 年 5 月，基础外墙钢筋施工

5 月

2015 年 5 月 20 日，基础工程分步验收

2015 年 5 月，首节钢柱安装完成

2015 年 5 月，土方回填开始

2015 年 5 月，第一节钢柱地脚螺栓安装完成

2015 年 5 月，土方回填完成

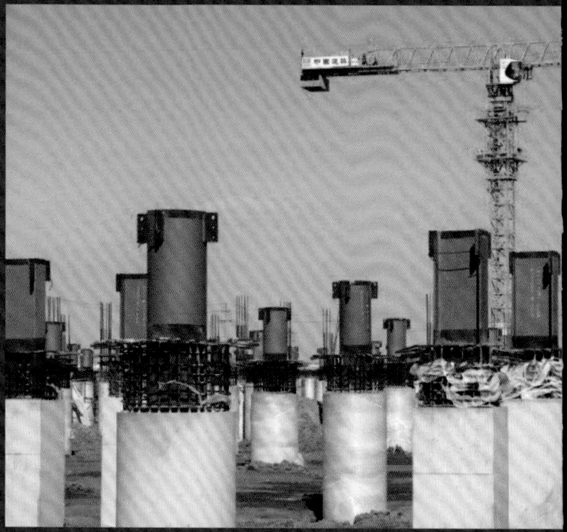

2015 年 6 月，地下钢柱外包混凝土完成

5 月

6 月

2015 年 6 月，底板防水卷材施工

2015 年 6 月，地下外墙混凝土完成

2015 年 7 月，零层板钢筋绑扎施工

7 月

2015 年 6 月，外墙挤塑板施工

2015 年 7 月，零层梁板混凝土施工

2015 年 8 月 23 日，地上钢结构开始吊装

7 月

8 月

2015 年 7 月，零层梁钢筋绑扎施工

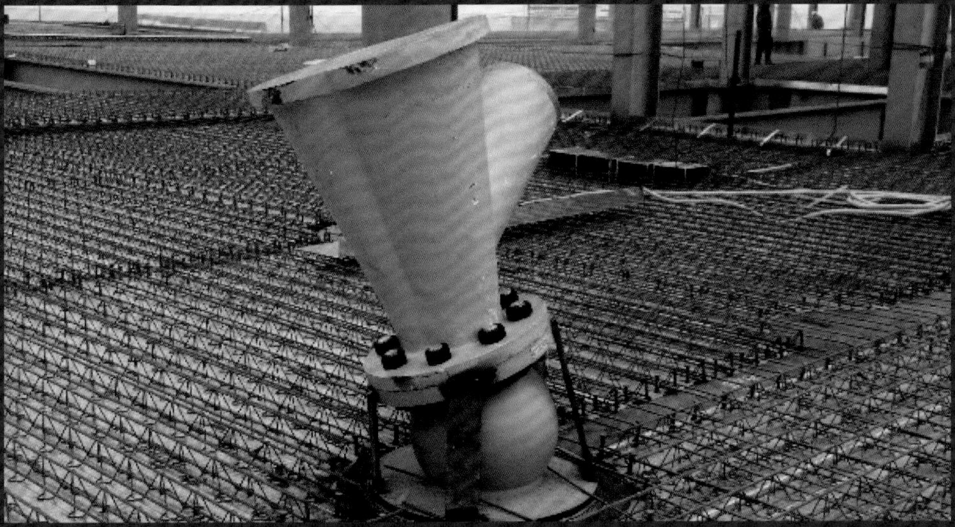

2015 年 8 月，抗震球形支座开始安装

2015 年 8 月，首层斜钢柱安装完成

2015 年 8 月，柱吊装对接

2015 年 10 月，安装悬挑端钢结构

9 月　10 月　11 月　12 月

2015 年 9 月，部分钢结构安装完成

2015 年 11 月，开始吊装门式钢桁架

2015 年 12 月，安装门式钢桁架

2015 年 12 月，抗震球形支座安装完成

2016 年 2 月，屋顶部分钢结构合拢

12 月　**2016 年**　**2 月**

2015 年 12 月，安装门式钢桁架

2016 年 2 月，安装钢桁架间斜撑

2016 年 3 月，吊装中央大厅首榀鱼腹式桁架

2016 年 3 月，安装中央大厅鱼腹式桁架

3 月

2016 年 3 月，安装幕墙檩条

2016 年 4 月，安装端部悬挑幕墙钢结构

2016 年 4 月，安装端部悬挑幕墙主钢桁架

4 月

5 月

2016 年 5 月，砌筑首层墙体

2016 年 5 月，中央大厅鱼腹式桁架防火涂料施工

6 月

2016 年 6 月，中央大厅安装幕墙檩条

2016 年 6 月，主体钢结构安装基本完工

2016 年 7 月，安装端部悬挑幕墙钢檩条

2016 年 8 月，安装幕墙外檐底板、T 码及屋面格栅

7 月

8 月

2016 年 7 月，施工钢结构防火涂料

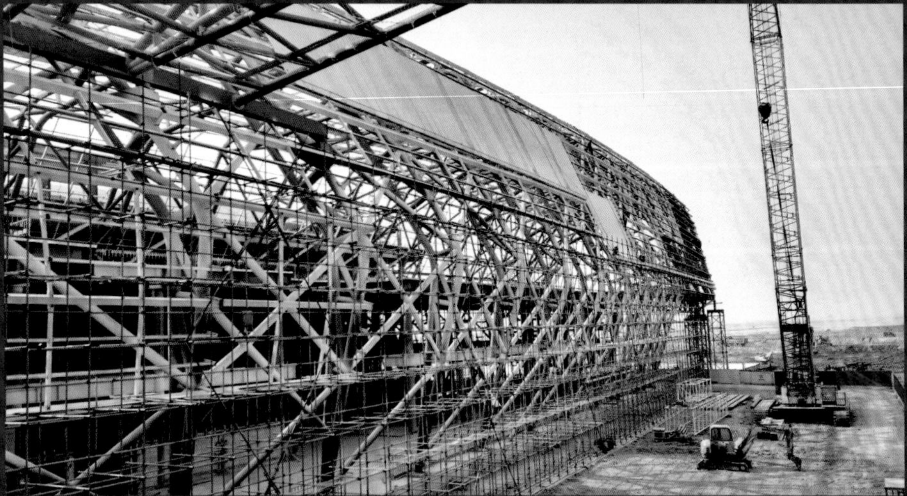

2016 年 8 月，安装端部悬挑幕墙钢檩条

2016 年 9 月，安装端部悬挑幕墙直立锁边板

9 月

2016 年 9 月，外檐幕墙岩棉板施工

2016 年 10 月，外檐幕墙直立锁边板施工

2016 年 10 月，玻璃幕墙施工

10 月

2016 年 10 月，安装悬挑端外遮阳板主钢结构

2016 年 10 月，外檐幕墙直立锁边板施工

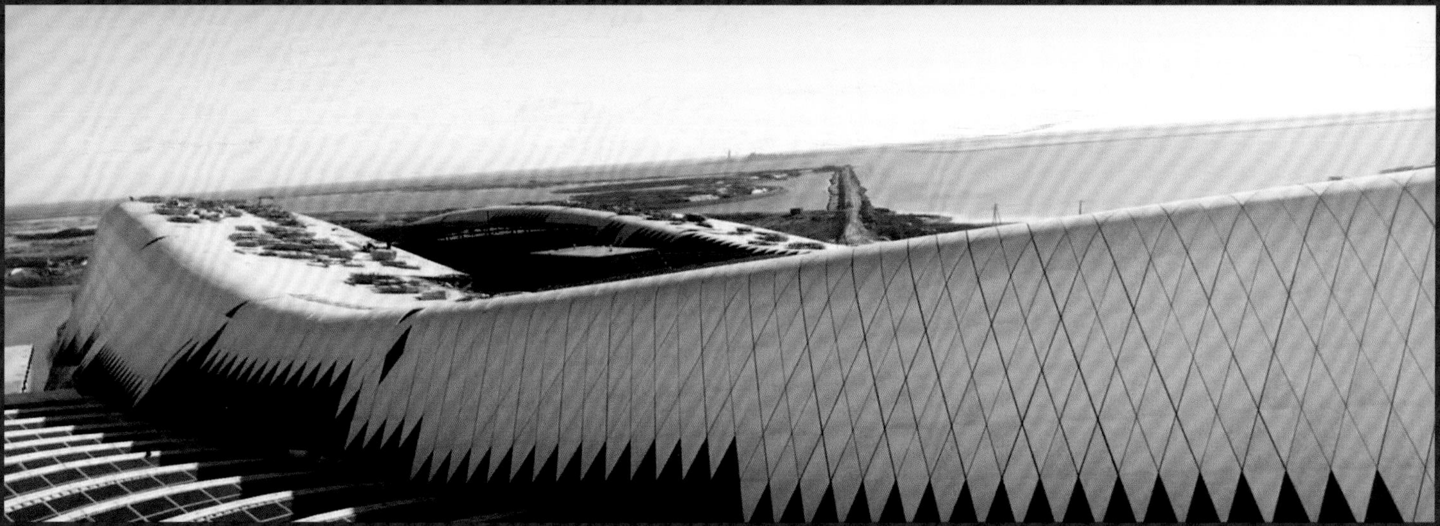

2016 年 10 月，中央大厅屋顶太阳能光伏屋面完成

11 月

2016 年 11 月，安装端部悬挑幕墙外墙板

2016 年 11 月，安装端部悬挑幕墙铝板龙骨

2017 年 1 月，外檐幕墙直立锁边板施工

2017 年 1 月，中央大厅屋顶幕墙施工

2017 年

1 月

2 月

2017 年 2 月，安装幕墙外饰板龙骨

2017 年 5 月，悬挑端头玻璃幕墙钢框架施工

3 月

4 月

5 月

2017 年 3 月，全部完成外檐直立锁边板

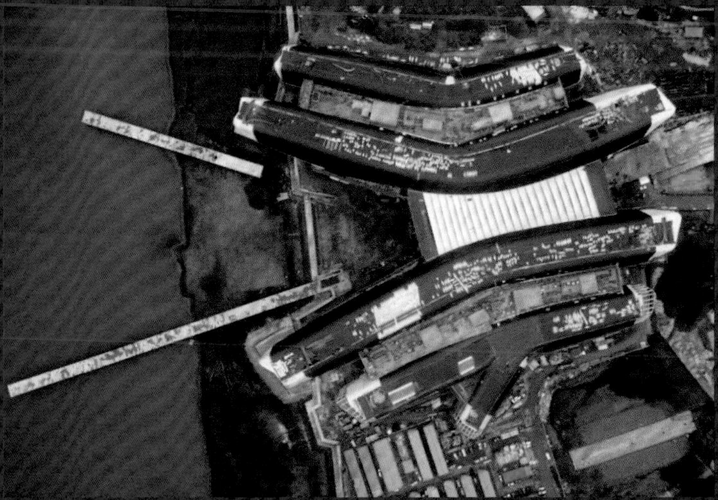

2017 年 4 月，外檐幕墙铝板开始施工

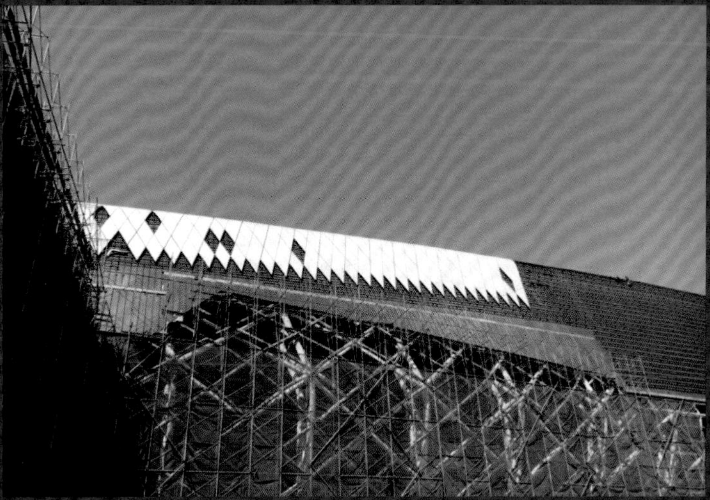

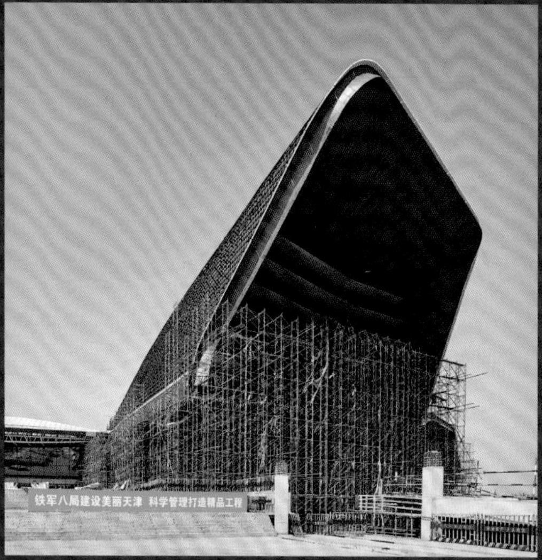

2017 年 6 月，安装端部悬挑幕墙内部木纹板

2017 年 6 月，外檐铝板施工完成 40%

5 月

6 月

2017 年 5 月，安装幕墙外饰板龙骨

2017 年 6 月，中央大厅内饰铝板施工

2017 年 7 月，幕墙遮阳铝板施工

2017 年 8 月，幕墙悬挑部位木纹板施工

7 月

8 月

9 月

2017 年 9 月，外檐铝板施工完成 60%

2017 年 10 月，外檐幕墙铝板开始施工

10 月

11 月

2018 年

2017 年 11 月，外檐石材、镜面铝板及玻璃幕墙施工

2018 年 3 月，室内精装修 GRG 施工

2018 年 10 月，大厅内部 GRG 施工完成

3 月

10 月

2018 年 10 月，中央大厅装饰完成

2018 年 10 月，幕墙屋面完成

2018 年 10 月，办公区屋顶幕墙格栅屋面完成

10 月

2018 年 10 月，中央大厅屋顶太阳能光伏屋面完成

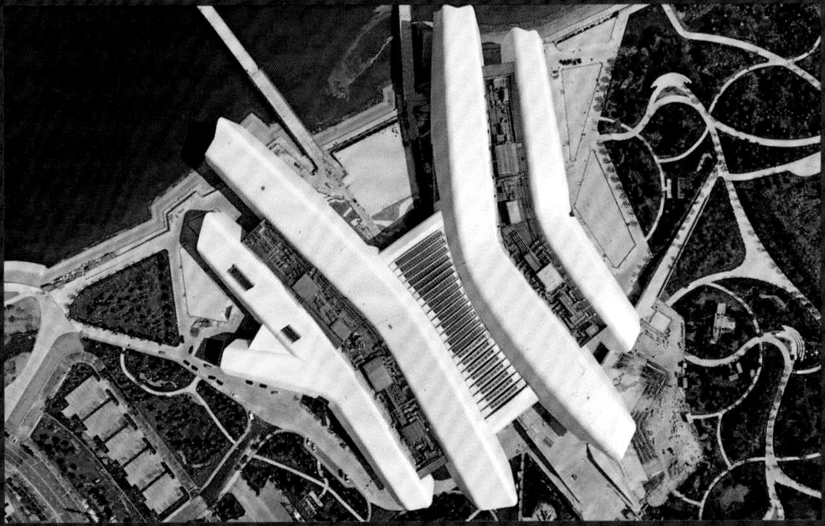

2019 年 3 月，外檐铝板施工完工

2019 年

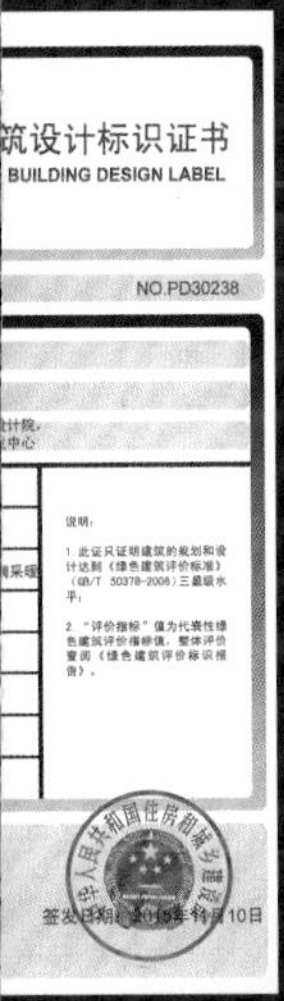

设计标识

AEC Excellence Awards（全球工程建设行业卓越奖）

2019 年度建筑设计类，中型项目组，最佳实践奖第一名

获奖证书

天津市建筑设计研究院有限公司：

你单位 国家海洋博物馆 被评为 2020 年“海河杯”天津市优秀勘察设计 建筑工程-公建 一等奖。

特发此证，以资鼓励。

天津市勘察设计协会

二〇二〇年九月

证书编号：2020-01-01-002

2020 年“海河杯”天津市优秀勘察设计一等奖

编后语：
书写国家“海博馆”创意设计的成长史

Postscript:
Writing the Growth Process of the Creative Design of the National Maritime Museum of China

金磊
Jin Lei

《中国建筑文化遗产》《建筑评论》编辑部是在2017年春天第一次走进建设中的天津 · 国家海洋博物馆的。因为工作原因，我们自认为对中外博物馆建筑有些了解，也于2010年受时任国家文物局局长单霁翔的委托，代表国家文物局为国际博物馆协会出版《中国博物馆建筑》（天津大学出版社，2010年11月第一版）。特别是近十年也跟踪了全国新涌现的有影响力的文博建筑的动态。但走进天津 · 国家海洋博物馆，让我们仿佛进入一种被打破沉寂的境界中，在刘景樑大师领衔下设计的天津 · 国家海洋博物馆这一新项目，实在令人眼前一亮。因为，一眼望去，它真是有世界领先水准的释放创意能力之作，它让每位走进它的人在感佩其艺术创作与科学严谨有机融合的功底时，也赞叹它是一介大师致敬美好、传承文明的“心 · 物”之作。

The editorial departments of *China Architectural Heritage* and *Architectural Reviews* paid their first visit to the National Maritime Museum of China in Tianjin under construction in the spring of 2017. Because of our work, we assumed that we knew something about Chinese and foreign museum architecture. In 2010, we were commissioned by Shan Jixiang, then Director of National Cultural Heritage Administration, to publish *Museum Buildings in China* (Tianjin University Press, November 2010, 1st edition) for the International Council of Museums on behalf of the National Cultural Heritage Administration. Especially in the recent decade, we have been also tracking the trends of the newly emerging influential cultural buildings in China. However, when we walked into the National Maritime Museum of China, it seemed that we were entering a realm where silence has been broken. The new project of the National Maritime Museum of Chinas designed under the leadership of Master Liu Jingliang is really eye-opening, because it is really a masterpiece featuring the release of creativity with a world-leading standard. It makes every visitor admire the solid foundation with the organic integration of artistic creation and scientific rigor, while marveling at it as a masterpiece paying tribute to beauty and cultural heritage.

记得看完天津·国家海洋博物馆回京当晚我就用微信致函刘景樑大师，对他创造的又一杰作表达祝贺之情。在接下来的2017年6月及9月，先后向刘大师递交了《天津·国家海洋博物馆》图书的编撰出版策划案，文案中强调该书必须达到能代表国家文博建筑水准并体现设计技术与文化内涵的品质。针对“天津·国家海洋博物馆”项目的设计生长足迹，以及国家海洋战略的发展部署，编辑团队初步认为该项目不仅是“文化天津”渤海之滨魅力之城的文化新地标，更应成为国家塑造中华海洋文明传承与创新的文化地标及国家形象，因此在几个版本的出版策划中都坚持了如下的关键语及目标。

● 在体现博物馆载体、利用“遗产日”文化与自然双重特点时，力求使该书成为有“国家馆”身份的范例文本。

● 在体现建筑设计与文博展陈完美结合共塑佳作时，力求使该书成为建筑师体味文博思想与需求的设计指导书。

● 在体现用建筑师的理念与设计技法营造海洋文明展示空间时，力求使该书成为国内首部用建筑文化对话中国海洋文化的著作。

● 在体现“人文海洋”的城市可持续发展理念时，力求使该书写就国内特有的体现生态文明“文海洋”绿色发展的实践个案图书。

● 在体现并记录建筑设计如何求索创新的方式与方法时，力求使该书在编制与文图创新上做到能成为有文创精神的全媒体出版物。

全国工程勘察设计大师刘景樑作为天津市建筑设计院有限公司（原天津市建筑设计院）名誉院长，五十多

I remember that I wrote a letter to Master Liu Jingliang on WeChat on the night when I returned to Beijing after visiting the National Maritime Museum of China in Tianjin, expressing my congratulations on another masterpiece he had created. Next, in June and September 2017, I submitted to Master Liu the compilation and publication plan of the book *Tianjin·National Maritime Museum of China*, which emphasized that the book must represent the national cultural architectural standards and embody the design technology and cultural connotation. In view of the design and growth footprint of the project as well as the development and deployment of the national maritime strategy, the editorial team had an initial understanding that the project is not only a new cultural landmark of "Cultural Tianjin", a wonderful city on the coast of the Bohai Sea, but also a cultural landmark and national image for shaping the inheritance and innovation of China's maritime civilization. Therefore, the following key words and goals have been abided by in several editions of the publishing plan.

● While displaying the museum carrier and making use of the dual characteristics of culture and nature of the "Heritage Day", we strive to make this book a model text with the status as a "national museum"。

● While reflecting the perfect combination of architectural design and museum exhibition, we strive to make this book a design guide for architects to appreciate cultural thoughts and needs.

● While showing the architects' ideas and design techniques used to create the exhibition space of a maritime civilization, we strive to make this book the first book in China featuring a dialogue between the architectural culture and the Chinese maritime culture.

● When embodying the concept of urban sustainable development of "cultural ocean", we strive to make this book a unique practical case in China that reflects the green development of the "cultural ocean" reflecting ecological civilization.

● When revealing and recording how to seek innovative ways and methods in architectural design, we strive to make this book an all-media publication with a cultural creative spirit in compilation and textual and graphic innovation.

As Honorary President of Tianjin Architectural Design Institute Co., Ltd. (formerly Tianjin Architectural Design Institute), Liu Jingliang, a National Master of Engineering Survey and Design, has worked hard in the front line of architectural design for over 50 years. In recent dozen years alone, he has directed and organized

年兢兢业业奋斗在建筑设计第一线，仅近十几年来先后指导、主持过在天津乃至全国颇具影响力的天津三大文化工程，犹如“三大战役”。我们也记不清为了解这些项目多少次采访过刘大师，但每次都能从他的忙碌身影及交谈中学到、感悟到许多，他不停地在为打造这些有生命力的文化地标默默耕耘。有人说，建筑艺术是心灵的跳动，从 2010 年的天津市文化中心“六大馆”建设组织与协调，到 2017 年主持、整合“五大馆”更繁复的天津·滨海文化中心的落成，再到 2019 年 5 月天津·国家海洋博物馆的部分对外开放，在国内外赞誉不断及观者无数掌声与欢颜中。刘景樑大师及其团队回报社会的是服务社会的使命与责任，他们的谦逊让我们更加理解这是成就一项项设计精品之关键，这里有他们团队拥有的优良创作土壤，也有守望初心的设计底气。《中国建筑文化遗产》《建筑评论》编辑部很荣幸在刘景樑大师的充分信任下，于 2018 年 11 月完成了天津“三大文化工程”之《天津 · 滨海文化中心》的编撰出版任务，如果说那本书是用时代镜像感悟滨海文化中心多机构联合设计的魅力之作，那么今天编撰完成的《天津 · 国家海洋博物馆》一书，将更加扎实且光彩照人，因为它彰显的是世界视野下，中国向海而兴且迈向深蓝的服务海洋文明传播的新力作。

1901 年，泛美博览会的导览序言中提到：“请记住，当你一步跨进门，你就已是被展示的一部分了。”充满创意的天津 · 国家海洋博物馆恰是这样的作品：它馆园结合、建筑与景观及场所的整体营造，体现了“馆舍天地”与“大千世界”的天然融合；建筑外形似跃向海面的鱼群及停靠岸边的欲起航的船，是自然海洋生态的舒展，更似建筑创作的自由挥洒，有形式的自由、尺度的自由、

three influential cultural projects in Tianjin, which have impact even on the whole country, just like the “Three Great Campaigns”. We cannot count how many times we have interviewed Master Liu to understand these projects, but we can learn a great deal from his busy work and from conversations with him every time. The master has been working quietly to create these living cultural landmarks. It has been remarked that architectural art is the beating of the soul. From the organization and coordination of the “four major halls” of the Tianjin Cultural Center in 2010, to the completion of Tianjin Binhai Cultural Center, which integrates the “five major halls” in 2017, and then to the opening of part of the National Maritime Museum of China in October 2019, amidst the countless honors at home and abroad as well as applauses and wonders of visitors.

Master Liu Jingliang and his team repay the society with the mission and responsibility of serving the society. Their humility makes us understand that this is the key to achieving excellent design products, on which resides the excellent creative soil owned by their team and also the design spirit of remembering initial aspirations. The editorial departments of *China Architectural Heritage* and *Architectural Reviews* were greatly honored, with the full trust of Master Liu Jingliang, to complete the compilation and publication task of Tianjin Binhai Cultural Center, one of the “three major cultural projects” in Tianjin, in November 2018. If that book is a work of understanding the wonders of multi-agency joint design of the Binhai Cultural Center with the mirror image of the times, then the book *Tianjin·National Maritime Museum of China* compiled today will be even more solid and brilliant, as it presents a new masterpiece serving the communications of the maritime civilization in China when the country is exploring the ocean and going into the deep blue.

In the preface of the guide to the 1901 Pan-American Expo, it is mentioned that “remember, when you take a step into the door, you are already a part of the exhibition.” The National Maritime Museum of China in Tianjin is just such a creative work: the museum-park integration, and the overall construction of architecture, landscape and place reflect the natural integration of “the world of the museum” and “the wide world”. The building is shaped like a school of fish jumping to the sea and a ship berthing on the shore about to set sail, which is the extension of natural maritime ecology, more like the free expressions of architectural creation featuring freedom of form, scale and environment. The combination of form and spirit and the unique architectural style result from the combination of contemporary design and technology,

环境的自由；建筑的形神合度、别样风采是当代设计与科技结合之果，是科学与人文在这高品质建筑中的交织之果，它是先进理念与新科技支撑的设计实践；天津·国家海洋博物馆在建筑师与文博专家日复一日的编织下，仿佛一个从海上飘来的“城”，在这里可领略到室内外柔美建筑力与美语境的不凡建构，更可为爱遐想的中小学生带来海洋文化与美育素养的体验式浇灌，无疑是公众走向海洋科技文化的精美课堂。

《中国建筑文化遗产》《建筑评论》编辑部从策划到投入承编《天津·国家海洋博物馆》一书历时两年，之所以今天能将满意的成书献给社会，得益于刘景樑大师领衔团队的精致作品，得益于建筑摄影师及美术编辑的图片艺术创作，也得益于国家对文旅深度融合的支持从而给建筑作品高质量发展注入了强劲动力。今天，我们站在新的历史交汇点上，希望所编撰这本有海洋“博物学”空间形态的建筑图书，不仅仅是忠实的记录设计历史并体现运维，也回答设计大师的创作之路是如何养成的，从而成为推动中青年建筑师设计成长的有用之书。有鉴于此，《中国建筑文化遗产》《建筑评论》同人，由衷感到我们的种种努力，是为传播当代中国建筑文化做了有益的事，这是我们整个团队的荣光。

《中国建筑文化遗产》《建筑评论》编辑部

2020 年 10 月 30 日

the interweaving of science and culture in this high-quality building, a design practice supported by advanced concepts and new technologies. With the weaving of architects and cultural experts day after day, the National Maritime Museum of China in Tianjin is like a “city” floating from the sea, where you can appreciate the extraordinary construction of gentle architectural power and beauty indoors and outdoors. More importantly, it can bring experiential education of maritime culture and aesthetic literacy to primary and middle school students who love fantasy. This is undoubtedly a fine classroom for the public to approach maritime science, technology and culture.

It took two years for the editorial departments of *China Architectural Heritage* and *Architectural Reviews* to compile the book *Tianjin·National Maritime Museum of China* from planning to completion. Today, the book is to be successfully published, thanks to the masterpiece accomplished by the team spearheaded by Master Liu Jingliang, thanks to the graphic creation of architectural photographers and art editors, and also thanks to the support from the Chinese government for the deep integration of culture and tourism, thus vigorously motivating the high-quality development of architectural works. Today, standing at a new historical intersection, we hope that this architectural book presenting the spatial form of maritime “natural history” will not only faithfully record its design history, reflect its operation and maintenance, but also answer how the creative path of design masters has been developed, thus becoming a useful book for promoting the design growth of young and middle-aged architects. In view of this, the colleagues of *China Architectural Heritage* and *Architectural Reviews* sincerely feel that our team is greatly honored because all our efforts have borne fruits for spreading the contemporary Chinese architectural culture.

Editorial Departments of China Architectural Heritage and Architectural Reviews

October 30, 2020

《天津·国家海洋博物馆》编委会

主编单位　天津市建筑设计院
承编单位　《中国建筑文化遗产》编辑部
顾　　问　单霁翔　安来顺　马国馨　崔　愷　盖广生　马华山　董　琦　孟庆金　陈　卓　赵春贵　陈　克　关　键　宋汝棻　钱秀丽
主　　编　刘景樑
策　　划　金　磊
副 主 编　菲利普·考克斯　陈天泽　张　馥　王　健　张丽丽
编　　委　孙学华　康　方　赵炳君　刘　伟　胡　健　刘　静　孟令華　林咸村　李　磊　尚　旸　王　珊　张　曦　苗　淼　朱有恒　董晨曦
影像摄制　刘　东　张　卓　刘洪彬　中建八局
翻　　译　陈晓霜（中译语通科技股份有限公司）

天津·国家海洋博物馆项目设计

总体设计　天津市建筑设计院
建筑设计　天津市建筑设计院、澳大利亚 COX 建筑师事务所
展陈概念　加拿大洛德文化资源规划和管理公司、北京洛德国际文化发展有限公司
展陈规划　北京洛德文博装饰工程有限公司、金大陆文化产业（集团）有限公司、北京清尚建筑装饰工程有限公司、天禹文化集团有限公司、上海安鲁展览展示有限公司
室内设计　天津市建筑设计院、澳大利亚 COX 建筑师事务所
景观设计　天津市建筑设计院、砼人意匠建筑工程设计工作室
泛光照明设计　天津市建筑设计院、北京对棋照明设计有限公司
绿建设计　天津市建筑设计院
智能化设计　天津市建筑设计院
BIM 设计　天津市建筑设计院、澳大利亚 COX 建筑师事务所
造价咨询　天津市建筑设计院
声学设计　天津大学建筑学院
影院设计　天津市建筑设计院、天津市舞台研究所
导识系统设计　天津市建筑设计院、天津时空风尚广告艺术工程有限公司
幕墙设计咨询　天津科林斯建源幕墙咨询有限公司

天津市建筑设计院设计团队

设计总负责人　刘景樑（国家勘察设计大师、天津市建筑设计院名誉院长）
专家顾问团队　张家臣　刘祖玲　朱铁麟　黄兆纬　宋　彻　乐　慈　伍小亭　周国民　刘建华　杨政忠　王东林　侯建成
总体设计　陈天泽　张　馥　孙学华　赵炳君　康　方　刘　伟　张淑梅
建筑设计　陈天泽　张　馥　江　屏　李　晔　王　健　张　丹　韩　冰　杨　波　李维航
结构设计　孙学华　张桂茹　赵晨杰　赵　扬　贾林毅　姬　爽
机电设计　杨政忠　赵炳君　康　方　刘　伟　蔡廷国　常　邈　韩金旭　林　娜　肖厚蓬　陈　玮　赖光怡　邓　祺　白　勇
室内设计　刘　杰　刘　静　陈　平　刘　彬　穆启明
绿建/BIM　李宝鑫　王敬怡　芦　岩　董璐璐　刘小芳　向　敏　刘　欣　卢婉玫　冯　佳　刘　振　张昭午　吴玉璇　曹　禹
造价咨询　周志良　于　力　杨　铭　曹学文　王　艳　汪海兰　王　红

图书在版编目（CIP）数据

天津·国家海洋博物馆 / 刘景樑主编．-- 天津：天津大学出版社，2020.11
ISBN 978-7-5618-6843-0

Ⅰ．①天… Ⅱ．①刘… Ⅲ．①海洋—博物馆—建筑设计—概况—天津 Ⅳ．①TU242.5

中国版本图书馆CIP数据核字（2020）第240733号

Tianjin · Guojia Haiyang Bowuguan

策划编辑：金　磊　韩振平
责任编辑：郭　颖
装帧设计：朱有恒
技术编辑：王　硉　刘博超

出版发行　天津大学出版社
地　　址　天津市卫津路92号天津大学内（邮编：300072）
电　　话　发行部：022-27403647
网　　址　publish.tju.edu.cn
印　　刷　北京利丰雅高长城印刷有限公司
经　　销　全国各地新华书店
开　　本　235mm × 286mm
印　　张　19.75
字　　数　130千
版　　次　2020年11月第1版
印　　次　2020年11月第1次
定　　价　248.00元